T0215114

AUGMENTATION TECHNOLOGIES AND ARTIFICIAL INTELLIGENCE IN TECHNICAL COMMUNICATION

This book enables readers to interrogate the technical, rhetorical, theoretical, and socio-ethical challenges and opportunities involved in the development and adoption of augmentation technologies and artificial intelligence.

The core of our human experience and identity is forever affected by the rise of augmentation technologies that enhance human capability or productivity. These technologies can add cognitive, physical, sensory, and emotional enhancements to the body or environment. This book demonstrates the benefits, risks, and relevance of emerging augmentation technologies such as brain-computer interaction devices for cognitive enhancement; robots marketed to improve human social interaction; wearables that extend human senses, augment creative abilities, or overcome physical limitations; implantables that amplify intelligence or memory; and devices, AI generators, or algorithms for emotional augmentation. It allows scholars and professionals to understand the impact of these technologies, improve digital and AI literacy, and practice new methods for their design and adoption.

This book will be vital reading for students, scholars, and professionals in fields including technical communication, UX design, computer science, human factors, information technology, sociology of technology, and ethics. Artifacts and supplemental resources for research and teaching can be found at https://fabricofdigitallife.com and www.routledge.com/9781032263755.

Ann Hill Duin is Professor in the Department of Writing Studies at the University of Minnesota.

Isabel Pedersen is Professor in the Faculty of Social Science and Humanities at Ontario Tech University.

ATTW Series in Technical and Professional Communication

Michele Simmons and Lehua Ledbetter, Series Editors

For additional information on this series please visit www.routledge.com/ATTW-Series-in-Technical-and-Professional-Communication/book-series/ATTW, and for information on other Routledge titles visit www.routledge.com.

AUGMENTATION TECHNOLOGIES AND ARTIFICIAL INTELLIGENCE IN TECHNICAL COMMUNICATION

Designing Ethical Futures

Ann Hill Duin and Isabel Pedersen

Routledge
Taylor & Francis Group

NEW YORK AND LONDON

Designed cover image: blackred / Getty Images

First published 2023
by Routledge
605 Third Avenue, New York, NY 10158

and by Routledge
4 Park Square, Milton Park, Abingdon, Oxon, OX14 4RN

Routledge is an imprint of the Taylor & Francis Group, an informa business

Library of Congress Cataloging-in-Publication Data
Names: Duin, Ann Hill, author. | Pedersen, Isabel, author.
Title: Augmentation technologies and artificial intelligence in technical
communication : designing ethical futures / Ann Hill Duin, Isabel Pedersen.
Description: New York, NY : Routledge, [2023] |
Series: ATTW series in technical and professional communication, 2639-3085 |
Includes bibliographical references and index. |
Identifiers: LCCN 2022058375 (print) | LCCN 2022058376 (ebook) |
ISBN 9781032263762 (hbk) | ISBN 9781032263755 (pbk) |
ISBN 9781003288008 (ebk)
Subjects: LCSH: Communication of technical information—Moral and ethical
aspects. | Performance technology—Moral and ethical aspects. | Artificial
intelligence—Moral and ethical aspects. | Intelligent agents (Computer
software)—Moral and ethical aspects.
Classification: LCC T10.5 .D85 2023 (print) | LCC T10.5 (ebook) |
DDC 601/.4028563—dc23/eng/20230111
LC record available at https://lccn.loc.gov/2022058375
LC ebook record available at https://lccn.loc.gov/2022058376

ISBN: 978-1-032-26376-2 (hbk)
ISBN: 978-1-032-26375-5 (pbk)
ISBN: 978-1-003-28800-8 (ebk)

DOI: 10.4324/9781003288008

Typeset in Bembo
by codeMantra

Access the Support Material: www.routledge.com/9781032263755

To our parents who forged the future

CONTENTS

FIGURES

TABLES

ACKNOWLEDGMENTS

The development of *Augmentation Technologies and Artificial Intelligence in Technical Communication* evolved amid the dynamism of rapidly changing landscapes. We wrote this book amid turbulent times, including a global pandemic, a war in Europe, and cases of geopolitical unrest and polarization across several continents. We know that augmentation technologies emerge in and through the contexts that provoke them affecting the people involved through their lived experiences.

We are grateful for the work of Sharon Caldwell, senior archivist for Decimal Lab, who helped us with research and integrating book chapters with https://fabricofdigitallife.com in creative ways. We thank Matthew Stein for his research collaboration on augmentation technologies and sharing knowledge of assistive technologies. We thank James Harbeck for his skillful and thorough copyediting.

We thank Grant Elgersma for his research assistance and insightful connections with industry; Brendan Nolan for research work on enhancement technology; Taher Dehkharghanian for his research assistance on augmentation technology and brain-computer interaction using his medical background; and Peter Turk for his editorial commentary on early drafts.

We thank those at Routledge for incredible help throughout the proposal, peer review, development, and publication process: Tharon Howard, editor of the ATTW Book Series in Technical and Professional Communication during our initial work with Routledge; Brian Eschrich, Editor of Communication Studies at Routledge; Sean Daly, Editorial Assistant, Communication Studies; anonymous peer reviewers for offering excellent direction; and current ATTW Book Series in Technical and Professional Communication editors, Michele Simmons and Lehua Ledbetter.

We are grateful for the insightful foreword by Heidi A. McKee and James E. Porter; their ongoing study of rhetorical and ethical approaches to understanding human-machine writing positions, technical and professional communicators for the multiple roles needed to be prepared for human–AI teaming. We thank Pam Estes Brewer and Saul Carliner for their endorsements of this book and their leadership for technical and professional communication.

We want to thank our ongoing collaborators as we study how artificial intelligence creates new roles and tasks for technical and professional communicators: Jason Tham and Gustav Verhulsdonck for implications stemming from smart cities contexts; and Nupoor Ranade and Daniel Hocutt for insight on using AI to understand users. We are thankful for international research colleagues at the Digital Life Institute (https://www.digitallife.org/) who constantly inspire us.

Ann thanks her Writing Studies colleagues for ongoing encouragement to explore writing and technical communication futures: Laura Gurak for her theoretical insight regarding trust; Lee-Ann Kastman Breuch for her solid pedagogical research; Molly Kessler for connections to the rhetoric of health and medicine; Dan Card for connections to environmental communication; and Mary Lay Schuster for her groundbreaking work in technical communication. She thanks members of the Technical Communication Advisory Board at the University of Minnesota for the many visits about emerging technologies, and she thanks her PhD seminar students who addressed how disciplines and professions might understand augmentation technologies and build digital and AI literacy to foster ethical futures: Caitllin Baulch, Chen Chen, Asmita Ghimire, Aaron Kapinos, Robert Kreml, and Alison Obright.

Isabel thanks Andrew Iliadis, Nawal Ammar, Susan Cody, Jay Cooper, Jack Narine, R. Bruce Elder, Patrick Hung, Ihor Junyk, Tanner Mirrlees, and Andrea Slane, whose rich collaborations inspire. She also thanks her graduate students from her Global AI Ethics course in the Master of Science (MSc) and PhD programs in Computer Science at Ontario Tech University who consistently challenge her to think about AI in new ways.

Some of our chapters drew on our previous articles. Table 2.2 Themes Associated with Transhumanist Rhetoric, ©2017 IEEE was reprinted, with permission, from Pedersen, I., & Mirrlees, T. (2017). Exoskeletons, transhumanism, and culture: Performing superhuman feats. *IEEE Technology & Society Magazine, 36(1),* 37–45.

Chapter 2 expanded and enriched our previous article, ©2017 IEEE was reprinted, with permission, from Isabel Pedersen and Ann Hill Duin, Defining a classification system for augmentation technology in socio-technical terms, IEEE International Symposium on Technology and Society (ISTAS) 2021.

Portions of this book and ideas were drawn from previous conference papers, including Isabel Pedersen and Ann Hill Duin, *AI Agents, Humans and Untangling the Marketing of Artificial Intelligence in Learning Environments* from the Proceedings of the 55th Hawaii International Conference on System Sciences, 2022. Chapter 7 drew upon ideas from Ann Hill Duin and Isabel Pedersen, *Working Alongside Non-Human Agents,* IEEE International Professional Communication Conference (ProComm) 2021 and Ann Hill Duin and Isabel Pedersen, *Tracing the Turn to Artificial Human and Human Teaming,* IEEE International Professional Communication Conference (ProComm) 2022.

Most importantly, we thank our family members for always supporting us, providing space, structure, and stimulating conversations about the future. We thank each other for another truly remarkable collaborative experience.

Last, we acknowledge that this research was undertaken, in part, thanks to funding from the Canada Research Chairs program and from the U.S. Council for Programs in Technical and Scientific Communication.

FOREWORD

Heidi A. McKee and James E. Porter

Miami University

> We proposed the book for engagement with a dynamic phenomenon, an emerging technology that operates more like a fast-moving train than its precursor, the traditional internet born in the 1990s that took 25 years to mature.
>
> — Ann Hill Duin & Isabel Pedersen, Chapter 8

In studying augmentation and artificial intelligence technologies, we are indeed boarding a fast-moving train. But with co-authors Ann Hill Duin and Isabel Pedersen we have marvelous guides who direct us through a bewildering array of developments and provide heuristic frames and ethical lenses for helping technical and professional communicators enter the rapidly changing landscape of human-machine interaction and collaboration.

In the title of this book—indeed, in the entire book—Duin and Pedersen raise a significant challenge to the field of technical/professional communication (TPC). Pedersen and Duin describe the development of augmentation technologies, articulate how responsible, ethical design provides the guiding principles for this development, and invoke TPC as the field that needs to be at the forefront of this development. The book signals a paradigm shift for technical/professional communicators, and provides us with a comprehensive critical framework, a much-needed road map, leading us there.

How a paradigm shift? Duin and Pedersen show how augmentation and AI technologies represent a leap, not a gradual transition, into a different kind of communication realm, one in which digital technologies merge with our personal experience. The digital communication technologies of the late 20th and early 21st centuries may have *assisted* us with our communications—but emerging AI-based writing systems can, increasingly, simply *do it for us*: they can handle communications autonomously, or nearly so, with minimal human guidance/prompting. They can perform much of the intellectual heavy lifting for communication.

But the change is not simply that AI systems can now communicate *for us*—write our reports, handle our customer service, create our marketing campaigns, produce our training videos. The even more dramatic shift is that AI systems are going to *become* us, and we them, as many of these technologies involve integration with the human body. In 1984, Donna Haraway noted that "We are all cyborgs"—and now Duin and Pedersen show us how that is becoming literally and technologically true. Augmentation technologies are embedded into our very beings, becoming integrated with our physical, emotional, and cognitive processes, shaping brain-computer interaction, amplifying memory, enhancing communicative abilities, creating extended, augmented, and virtual realities, and shaping social and rhetorical relationships via robots, virtual assistants, and digital humans.

Of course, we have had augmented humans for a long time: since the invention of sensory enhancements like eyeglasses, binoculars, microscopes, hearing aids, and even before that with the cane (to assist walking) and the ear trumpet (to assist hearing). Whenever we pick up a pencil to write, we are augmenting ourselves—but augmentation technologies refer to something substantially different. Augmentation technologies do not merely assist the body, like a cane or pencil, but they become part of the body—and not just prosthetically, but sometimes even neurally by adding to and changing our emotional and cognitive capacities. The pencil does not change our brains (too much), it doesn't mess with our cognitive or emotional functioning. But a writing system that links our brains directly, through BCI technology (brain-computer interface), to an AI-based smart robot, allowing for real-time mental collaboration, well, we are way beyond tools at that point. The train has ventured into territory that once was the exclusive province of science fiction.

Duin and Pedersen's insight, one of many, is to frame these augmentation technologies as fundamentally *communication* technologies. These technologies do not always produce words and images, but they communicate through interaction—human-machine interaction and, increasingly, machine-machine interaction. Thus, our fundamental notions of literacy need radical updating. What does literacy mean when people receive sensory and communicative input from their clothing, and when their clothing then communicates, autonomously, for them? What happens when a technical report is substantially produced by an AI system that not only makes content decisions about what data to include/exclude but also designs data visualizations and provides strategic recommendations for executive action? What happens when human technical communicators are working alongside, teaming with, AI-based digital employees or even smart robots who stand there chatting with us at the water cooler? (Notice the use of *who*.) Or, even, as Duin and Pedersen discuss at length with their fascinating chapter on digital employees, what happens when we cannot tell the difference between our human and AI colleagues. The field of TPC must prepare for these developments—and quickly, because they are not coming in the future, they are already here.

This book provides a rich tour of augmentation technologies and AI, bringing together global and interdisciplinary perspectives to understand AI-based human-machine collaboration and experience. Throughout the book are links to materials posted to the *Fabric of Digital Life* (https://fabricofdigitallife.com), an online archive of materials curated with comprehensive meta-tagging, supporting searches based on (to name just a few tags) discursive types, media types, relation to body, technologies, marketing, related entities/companies, and HCI type (including ambient, carryable, embeddable, ingestible, wearable, implantable, robotical, and other). These sample materials—e.g., the videos showing Sophia in action, the interview with the designer of bionic contact lenses, the research article on clothing and haptic action—are an important supplement to the book, providing a comprehensive archive for understanding the technology as it changes and advances. Someday, and someday very soon with the pace of technological change, the technologies discussed in this book will not be present and future, but past. And in this regard, Duin and Pedersen's work in the book and in the *Fabric* functions as historical recordkeeping, capturing trends and developments from the late teens and early twenties of the current millennium. The array of material they assemble and the fields from which they draw—media, communications, medicine, engineering, etc.—serve as an important archive for future research.

Why board this train? And, more importantly, how may we influence the directions augmentation and AI technologies are heading? One of the key contributions of the book is to highlight the importance of ethics, and ethical design, as providing the guiding principles for theory, research, and practice. We are on this train for a purpose—to design ethical futures. And to that end, Duin and Pedersen provide key lenses and frames for doing so. For educators, they provide a rhetorical literacy frame—to help us consider how we might engage ourselves and our students with developing the literacies needed to analyze, evaluate, interact with, and design digital augmentation technologies. Their ethical framework draws from multiple fields and many organizations, including (to name just a few) the World Intellectual Property Organization, the Berkman Klein Center at Harvard University, the AI Now Institute, the Institute of Electrical and Electronics Engineers (IEEE), and the Fabric of Digital Life by the Decimal Lab at Ontario Tech University.

The authors' view of ethics is an expansive one, aligned with TPC principles such as user-centered design, social justice design, and rhetorically situated design—but expanding beyond these topics to lay out a program for technological oversight and intervention. Technical/professional communicators who work, teach, research in the AI sector—and that will be all of us, eventually—need to practice responsible design, and that requires taking an expansive view of the implications of design practice, noticing at every turn its ethical consequences. Chapter 5, especially, does an impressive job of laying out the scope and importance of these ethical responsibilities—and also showing how we can be proactive and practical in realizing these principles in meaningful action. Duin and Pedersen call on us to approach the design and use of augmentation and AI technologies within the ethical considerations of privacy, accountability, safety and security, transparency and explainability, fairness and non-discrimination, human control of technology, professional responsibility, and the promotion of human values.

Overall, this book brings three main concepts together—*technology, ethics,* and *technical communication*—as a necessary set of interrelated concerns. Augmented and AI-based technologies are creating dramatic change in every area of human experience, which in turn require the attention of technical/professional communicators, who are the acknowledged communication experts. These experts need to be guided by principles of ethical design and committed to the practical, strategic actions needed to make sure that communication technologies meet human needs, reduce harm and suffering, and instantiate the values of justice, equity, privacy, and access. The ultimate goal is to "design ethical futures." A tall order indeed—and that is the provocative and important challenge this book presents to the field.

AUGMENTATION TECHN
AND AI—AI ETHICAL DESIGN
FUTU

SECTION ONE
Understand (Rhetorics of) Augmentation Technologies

Section One introduces the Ethical Futures Framework, defines and articulates the dimensions of augmentation technologies, focuses on the development of AI agents in industry as a critical factor in the rise of augmentation technologies, and provides critical attention to hyped assumptions and recent claims concerning the Metaverse and AR.

DOI: 10.4324/9781003288008-1

1
AUGMENTATION TECHNOLOGIES AND AI—AN ETHICAL DESIGN FUTURES FRAMEWORK

Overview

Augmentation technologies, fueled by artificial intelligence (AI), are undergoing a process of adaptation and normalization geared to everyday users in various roles as practitioners, educators, and students. While new innovations, applications, and algorithms are developed as "augmentation technology," Chapter 1 focuses on human subjects, contexts, and rhetorical strategies proposed for them by external actors. The chapter discusses core functions of technical and professional communication and provides rationale for positioning technical and professional communicators (TPCs) to understand augmentation technologies and AI as a means to design ethical futures across this work. An overview of *Augmentation Technologies and AI—An Ethical Design Futures Framework* serves as a guide for reframing professional practice and pedagogy to promote digital and AI literacy surrounding the ethical design, adoption, and adaptation of augmentation technologies. The chapter concludes with an overview of the remaining chapters in this book.

Key Questions

- How might augmentation technologies best be defined?
- What powerful rhetorics signal the augmentation technology and AI promise of augmented humans, cyborgs, and posthumans?
- The progress of augmentation technologies and AI follows decades of theoretical research on cyborgs, posthumanism, transhumanism, and non-humanism in philosophy, critical theory, critical feminism, and feminist new materialism. How might technical and professional

DOI: 10.4324/9781003288008-2

communicators use this research for building understanding of augmentation technologies and AI?

- How will core functions of technical and professional communication—understanding audience, user experience, content development, collaboration, and design—transform alongside augmentation technologies and AI?

Chapter 1 Links

Throughout the chapter, we refer to articles, videos, and reports which can be found in a related chapter collection at Fabric of Digital Life called *Augmentation Technologies and AI—An Ethical Design Futures Framework* (Duin & Pedersen, 2023). You can find a link to this collection at https://fabricofdigitallife.com.

Introduction

The Press Release for Gartner Inc.'s Hype Cycle for Emerging Technologies (2020) describes 30 "must-watch" technologies. Amid descriptions for changing virtual workplaces and solutions for the global pandemic crisis appears a seemingly mundane reference to technologies that will "alter the state of your brain." The release explains, "the way people interact with the digital world is also moving beyond screens and keyboards to use a combination of interaction modalities (e.g. voice, vision, gesture), and even directly altering our brains." Chris Duffey (2019), writing in preparation for the CES 2020 Las Vegas exhibition, asks, "What if we could tap into superhuman powers to be better at school, excel at sports, succeed in business and ultimately live a longer and fuller life? Who wouldn't want that competitive edge?" And corporate innovation announcements reflect this trend: "Elon Musk says Neuralink could start planting computer chips in human brains within the year," emphasizing the trials that will be conducted to shift brain implants from a speculative proposed technology to one ready for adoption (Kay, 2021).

The purpose of this book is to cultivate an even deeper understanding of human augmentation and AI technology, build technical and professional communication capacity to articulate its benefits and risks, and provide direction for future practice and collaboration. This book is concerned with augmentation technologies and AI that will alter the brain, but it is also concerned with recent discourses that are so ready to promote brain augmentation as a paradigm shift. Katina Michael and colleagues (2020), in their review of the rise of implantables, emphasize that "in some ways, the end user is the new 'last mile' in the global interconnected network topology" (p. 97). With this point, we argue that augmentation technologies fueled by AI are undergoing a process of normalization geared to everyday users in various roles as practitioners, caregivers, educators, and students.

For example, regarding practitioners, a 2021 Massachusetts Institute of Technology (MIT) Sloan Management Review (SMR) survey, "The Workforce Ecosystem Perspective," suggests that the meaning of the term "workforce" is undergoing a rapid shift, due in part to emerging augmentation technologies. Of the 5,118 global executives surveyed, MIT SMR found that 87% of respondents "consider their workforce to encompass more than their employees"; 37% view "technology for workforce augmentation" as "part of [their organization's] workforce"; and 56% indicate that this category will increase in the next two years. However, despite the trend toward increasing adoption of augmentation technologies and AI in the workplace, only a minority of these global executives believe their organization is adequately prepared to manage a future workforce made up of so many "external participants."

Given this workforce context, MIT SMR (2022) proposes that executives replace the traditional employee framework with "a workforce ecosystem approach." This would require new management strategies and a reconsideration of underlying philosophies, systems, and processes. They provide specific guidance for workforce planning, talent acquisition, performance management, compensation and rewards, learning and development, career paths, and organizational design. Similar to how MIT SMR addresses the need to adapt management approaches to the changes that augmentation technologies bring to organizations, technical and professional communication practitioners, instructors, and researchers will need to adapt to similar changes currently being normalized.

Part of this process of normalization includes narratives that explore ambiguities tightly bound with the pursuit of augmentation. *Black Mirror*, created by Charlie Brooker (2011–2019), is a popular science fiction television series that deals with augmentation in stark dystopian terms. Reminiscent of the near future worlds of *Ex Machina* and *Her*, *Black Mirror* often concentrates on the harsh consequences of implanting humans with computer chips. In season one, episode 3, "The Entire History of You," characters have a "grain" implanted behind their ears that records everything they see and hear. Using a remote, they can play back memories on screens. It illustrates an enduring real-life ambition for technologies to augment human memories by preserving all in digital archives. In another episode, a doctor receives a neurological implant that allows him to feel the physical pain of his patients, thus being able to diagnose them more accurately through an augmented sense of empathy.

Black Mirror stands in contrast to the celebratory futurism of Marvel's cinematic universe franchise of superhero films or even the uber-cool urbanism of the *Matrix* trilogy (Wachowski & Wachowski, 1999–2003). It critiques tactics of current technology adoption based on providing work efficiencies that replace human work or activities with autonomous agents. It sometimes sets the narratives in raw natural landscapes, emphasizing the non–digital materiality of life and uncontrollable environments that resist human technological interference. *Black Mirror* emphasizes socio-political contexts that position augmentation technologies as instigators of inequalities in society—the digital divide. Relevant as well is the way it frames its narratives on current innovations in research and development. Empathetic AI and augmenting human empathy are two real-world technologies under development (Dial, 2018; Wu, 2019): Rachel by Emoshape uses AI algorithms to respond to meaning and language around her in real time, then expresses herself accordingly (Emoshape Inc., 2018). *Black Mirror* not only critiques current emerging human augmentation technologies, it also scrutinizes Big Tech's interest in this sector as well as mainstream media's recent preoccupation with reporting on it.

If truth is stranger than fiction, there is no better example than the dramatic rise of AI generators in 2022 and 2023 instigated by Big Tech corporations fueling hype about AI at an astonishing rate. OpenAI's ChatGPT chatbot is an AI text generator used by millions of people for writing and augmenting cognitive work, yet its rapid adoption over the course of a few weeks has caused people to fear its implications. AI text generators are often referred to as dangerous and powerful due to their ability to mimic human communication across multiple writing genres. Likewise, AI image generators like Midjourney, OpenAI's DALLE-E, and Stable Diffusion are disrupting all graphic design fields.

The discourses that justify augmentation technologies use powerful rhetorics promising to deliver transparency in AI decision-making along with augmented humans, cyborgs, and posthumans as a tactic for adoption. We also acknowledge that many individuals self-identify as augmented or aspire to augment themselves using technologies, including those who have garnered some kind of fame (each of the following is included in the Fabric of Digital Life collection, *Augmentation Technologies and AI—An Ethical Design Futures Framework*): Stelarc (n.d.); Rob Spence (Spence & Jaworski, n.d.); and cyborg activists Neil Harbisson (n.d.), Kevin Warwick (Infonomia, 2008), and Moon Ribas (n.d.). Their autobiographical narratives weave throughout discourses about technologies. Biohacking, the lifestyle practice of using science and technology to improve one's physicality or cognitive abilities, thus becomes more mainstream. Biohackers are discussed in popular science venues, popularized by Silicon Valley titans. One *Vox* article describes it as "one branch of transhumanism, a movement that holds that human beings can and should use technology to augment and evolve our species" (Samuel, 2019).

Augmented Humans: Histories, Theories, and Rhetorics

The progress of augmentation technologies follows decades of relevant theoretical research on cyborgs, posthumanism, transhumanism, and non-humanism in philosophy, critical theory, critical feminism, and feminist new materialism (Haraway, 1985; Hayles, 1999; Barad, 2003; Wolfe, 2010; Braidotti, 2013). Transhumanism challenges bioethical positions due to its encouragement of human enhancement as a concept, a moral stance that continues to reverberate across augmentation discourse.

Theories of cognitive augmentation are often rooted in cybernetics theory. Influenced by Maurice Merleau-Ponty, Andy Clark proposed the "extended mind" theory in 1997, leading to a tradition that is in many ways key to augmentation technology strategies today. Clark writes of evolution as:

> a process that must build new solutions and adaptive strategies on the basis of existing hardware and cognitive resources. And it is empowered, as we have seen, by the availability of a real-world arena that allows us to exploit other agents, to actively seek useful inputs, to transform our computational tasks, and to offload acquired knowledge into the world.
>
> *(1997, p. 88)*

In later work he offers a nuanced take on extended mind to predict progress toward cyborgism:

> We shall be cyborgs not in the merely superficial sense of combining flesh and wires but in the more profound sense of being human-technology symbionts: thinking and reasoning systems whose minds and selves are spread across biological brain and nonbiological circuitry.
>
> *(Clark, 2003, p. 3)*

N. Katherine Hayles (2002),. prominent posthuman theorist, comments on Clark's reasoning and states, "Agency still exists, but for the posthuman it becomes a distributed function" (p. 319). She clarifies the subject's lived experience: "Living in a technologically engineered and information-rich environment brings with it associated shifts in habits, postures, enactments, perceptions—in short, changes in the experiences that constitute the dynamic lifeworld we inhabit as embodied creatures" (p. 299). Throughout this book, we draw on these points of view to explain augmentation technologies across various forms of enhancement, as well as directions to frame ways to presuppose their emergence across the platforms now under development and being used.

Historically, enthusiast groups and organizations connected to human enhancement have driven popular and oftentimes controversial momentum for augmentation as a positive, ambitious goal for humanity. These movements contribute terminology and motivate people to consider issues that might threaten people's value systems or even legal systems. Radical life extension, explored by Ray Kurzweil and Terry Grossman (2009), has pushed ethical boundaries for some, impacting the uptake of certain technologies. Groups like Humanity+ (also known as the World Transhumanist Association), Extropy Institute (no longer active), and Singularity University feature speakers, scientists, researchers, lawmakers, and thought leaders, according to each group's agenda. This book does not focus on or promote these groups or these movements.

Integrated with philosophical and existential categories for human-machine mergers important to the idea of *augmenting* humans are rhetorical interventions that cause people to adopt or reject them. Also important to this book is Kevin Thayer's (2014) work "Mapping Human Enhancement Rhetoric," in which he ponders how "questions can be raised about the shifting ethical positions in human and transhuman enhancement discourse" (p. 31). Gregory Hansell and William Grassie (2011) further identify transhumanism as a value system or belief structure with an ideological foundation trusting and expecting that dramatic human enhancement is possible:

> The applied sciences involved include dramatic advancements in the neurosciences, genomics, robotics, nanotechnology, computers and artificial intelligence. In some combination of the above bioengineering, transhumanists imagine the possibilities in the near future of dramatically enhancing human mental and physical capacities, slowing and reversing the aging process and controlling our emotional and mental states. The imagined future is a new age in which people will be freed from mental disease and physical decrepitude, able to consciously choose their 'natures' and those of their children. At first glimpse, it all seems like a wonderful thing, life lived more abundantly, but Francis Fukuyama calls this transhumanist vision 'the most dangerous idea in the world.'
>
> *(p. 13)*

One distinguisher of transhumanism is the value of human control over emotional and mental states. Human control is a contested value of current and proposed augmentation and one that we explore throughout this book in light of automation.

Following several of these premises, Isabel Pedersen and Tanner Mirrlees (2017) argue that "transhumanism reveals four predominant and often sensationalized themes used to promote human technological advancement as a seemingly obvious future in the mainstream" (p. 40). They argue that transhumanist rhetoric is usually buttressed with (1) a legitimate claim of technological or scientific progress; (2) a claim based on human agency or human control; (3) a claim for superheroism or a superhuman ability; and (4) an urgent claim based on a need or issue over human vulnerability. Using the example of exoskeleton development for physical, non-medical human enhancement, they conclude that "exciting news about technological innovations thrives across networked channels. People consume and circulate popular techno news as entertainment to tantalize, to scandalize, and to authorize, but as a result, society simultaneously accepts values in terms of this hyped technology" (p. 44).

Such public sensationalism along with fear become governing rhetorics of augmentation technologies. The development of a so-called superintelligence has fueled mainstream angst over the danger of evolving an Artificial General

Intelligence (AGI) for humans. This kind of public concern led Silicon Valley entrepreneur Elon Musk to perpetuate the myth that humans ought to enhance themselves to keep up with AI. An element of this book is the monitoring of such public justifications for technology under extremely hyped or even illogical conditions. In many ways the rhetoric of augmentation technology and AI is also technoliberal rhetoric:

> Technoliberal rhetoric is most visible as it is articulated to digital technologies, diffusing the terms, tropes, and frames of technoliberalism through sites of tech evangelism like Apple Keynotes, TED Talks, CES (the technology trade show formerly known as the Consumer Electronic Show) and through commercial advertisements across different media.
>
> *(Pfister & Yang, 2018, p. 241)*

We identify technoliberalism as "the condition of digitality [that] intensifies neoliberal governing rationalities in the context of public sphere deliberation…. Technoliberalism may be the default public philosophy of digital culture, but it is neither natural nor inevitable" (Pfister & Yang, 2018, p. 250).

Writing about communication, machines, and human augmentics, John Novak and colleagues (2016) ask, "When do electronic tools cease to be 'simply' tools, and become meaningfully part of ourselves?" From this pivotal question, they add "When might we think of these tools as augmenting ourselves, rather than simply amplifying our capabilities?" Echoing Mark Weiser's 1991 claim of ubiquitous computing, that society's inclination is to "push computers into the background," augmentics is contextualized in ontological terms. Robert Kenyon and Jason Leigh (2011) adopted the term "human augmentics" to describe "a call to arms for the rehabilitation [medical] community to think outside of their boundaries—to think of the problem in terms of a larger interconnected ecosystem of augmented humans rather than a patchwork of disconnected sub-systems" (p. 6761). Zizi Papacharissi (2019) uses the term "augmentics" to describe a contested state relating to expanded human capability: "By focusing on the theme of augmentics, I aimed at directing our attention to ways in which technologies expand our capabilities" (p. 7). Rejecting a state of assumed betterment, Papacharissi settles on difference, stating that:

> we will not become better, superior, or more advanced—that will be a function of how we put technology to use, and ultimately, a call that will be forever subjective. We will, or rather, we have the opportunity to, become different.
>
> *(p. 7)*

Useful too in Papacharissi's collection is Douglas Guibeault and Joel Finkelstein's (2019) interpretation of augmentics. Following Marshall McLuhan, they envision it within the cybernetic turn and use ecological principles to explore selves in decentralized contexts, articulating the possibility for an "augmented collective consciousness" (p. 179).

Leading rhetorical and professional communication scholars Heidi McKee and James Porter, in their 2017 chapter "AI Agents as Professional Communicators," acknowledge that "increasingly, humans are communicating with AI agents, often without knowing they are doing so. The implications of AI for professional communication and for organizations and business professionals who deploy AI agents are profound" (p. 135). In their exploration of rhetorical and ethical issues in AI within professional communication, their overarching question is "can AI bots be effective professional communicators?" (p. 135). Based on their case studies of use of AI agents, they articulate the most "interesting complexity of AI for rhetorical interaction" as occurring "when AI agents aim to become more 'intelligent' and when interactions become potentially even more open-ended, existing in the rhetorical free-for-all that is, ultimately, human conversation" (p. 153). Chapter 7 of this volume includes examples of digital employees designed for such communication.

Continuing with focus on rhetorical context for AI writing, in their 2020 work McKee and Porter emphasize that AI writing systems are "built on an information transfer model of communication that assumes text production is a simple matter of converting raw data into sentences and paragraphs" (p. 110), and that "When humans and AI systems interact, miscommunication occurs and ethical issues arise from lack of understanding about [social and rhetorical] context" (p. 113). They offer two ethical principles to guide design of AI writing systems, principles that we align with throughout this book:

- An ethic of transparency: humans must know the rhetorical context and whether they are interacting with an AI agent—whether in mobile text, social media, or other communication (p. 113); and
- An ethic of critical data awareness: a methodological reflexivity about rhetorical context and omissions in the data that need to be provided by a human agent or accounted for in machine learning (p. 110).

And in their 2022 study of "Team Roles & Rhetorical Intelligence in Human-Machine Writing," McKee and Porter continue to acknowledge the "immense effect" that fast-evolving AI writing technologies have on professional communication. While most current AI writing systems are "resource tools or assistants," they note that in "bounded verticals" AI writing agents function as "higher-contributing team members... who are... acting as *writers*" (p. 3). They again note the critical importance of understanding the rhetorical context for communication, with a call for considering "rhetorical intelligence" for AI. Moreover, they stress that "we are on the onset of a seismic change in writing and teaming for professional communication" (p. 390), with which we strongly agree. Critical to navigating this change is understanding of augmentation technologies, as both a rhetorical and ethical phenomenon.

Augmentation Technology Definition

Isabel Pedersen and Andrew Iliadis (2020) define "embodied computing" as those technologies that "exist in topographical [on the body], visceral [in the body], and ambient [around the body] relationships with the body" (p. xi). As a kind of embodied computing technology, augmentation technologies and AI enhance human capabilities or productivity by adding to the body (or ambient environment around the body) cognitive, physical, sensory, and/or emotional enhancements. Examples include brain–computer interaction devices for cognitive enhancement; robots marketed to improve human social interaction; wearables that extend human senses, augment creative abilities, or overcome physical limitations; implantables that amplify intelligence or memory; devices or algorithms for affective computing; Internet of Bodies (IoB) and Internet of Things (IoT) for ambient interaction or surveillance with places/spaces; and extended reality (XR) technologies, e.g., augmented reality (AR) and virtual reality (VR), to alter human interaction with people's lived reality. Judith Hurwitz and colleagues, longtime scholars of AI, write, "The most pragmatic and useful way to benefit from AI and machine learning is to implement these powerful technologies as an augmentation to human intelligence" (Hurwitz et al., 2020, p. xix).

Most importantly, we redefine augmentation technologies and AI as social and rhetorical phenomena. For example, social robots and virtual assistants are slated to augment human experiences and relationships, to communicate with, care for, monitor emotions of, entertain, instruct, and supervise humans, and to assist in teaching and practice. Human experiences, social life, arts and human identities, and the practice of technical and professional communication are affected by this momentum (Hartzog, 2015; Slane et al., 2020).

Augmentation technologies work to enhance the way we do things and perceive our environment. Therefore, human augmentation (Human 2.0), once assumed to focus solely on extending human capabilities, now works to transform human abilities for non-medical reasons. According to an expert panel from the Forbes Technology Council (2020):

> Human augmentation technology offers vast potential for many different industries. These developments combine medicine and technology to increase the capabilities of the human body. However, those outside the tech industry may not even be aware of how much augmentation can enhance our daily lives.

For example, Scribe.ai is working to:

> supplement our neurons to make even the most mundane occurrences into something unforgettable.... The plan is to methodically capture and store all sorts of data—audio, video, and eventually biometric—that can be easily searched or cleverly invoked in a way that augments your actual memory.

Scribe.ai's first product is an add-on to Zoom in which "a faceless participant... a dynamic rapporteur" would join a meeting to log "what people say and what they look like as they say it," which then might be used to augment memory and future work in more immersive ways (see Figures 1.1 and 1.2). This work is described in the article "A New Company Pursues Total Recall—Starting With Zoom," by Steven Levy (2021).

Augmentation technologies are alluring and celebrated. They are afforded the freedom to operate as if necessary and imminent rather than speculative because of their shocking facility to enhance human abilities and practices. At the same time, they evince a significant lack of forethought, governance, societal, or consumer control. Augmentation technologies emerge amid fluctuating corporate and international spheres that have not yet been regulated. For example, Moodmetric and Oura smart rings are two products designed with self-tracking capabilities and marketed with the promise of promoting wellbeing. In "Making sense with sensors: Self-tracking and the temporalities of wellbeing,"

FIGURE 1.1 Any participant may invite scribe to join the meeting, start recording, transcribe what is said, summarize key points, and search all past transcripts for key points. Used with permission of Dan Siroker, CEO of Scribe.

FIGURE 1.2 Scribe's user dashboard. Used with permission of Dan Siroker, CEO of Scribe.

Martin Berg (2017) analyzed rhetoric used to explain the functionality of these rings. In marketing materials and user manuals, Berg found language about the impenetrable nature of the body's natural signals, conceptions of the body as a machine that needs "optimizing," and assumptions about accelerated time in modern society.

Moodmetric and Oura are part of what researchers Mark Andrejevic and Mark Burdon (2014) call the sensor society. In a 2015 TED talk, Burdon discussed corporate rhetoric used to sell sensorized objects, but he also warned of ethical issues associated with these new products connected to large data collection systems. Now is the time, Burdon argues, to ask fundamental questions about how such devices can be designed and used ethically.

Consumers are already adopting the technology, and the industry is growing rapidly (Kundan et al., 2021). At CES 2022 in Las Vegas, many new digital health devices with sensor technology were on display. Abbott Laboratories, a medical device company, announced a push into the area of biowearables during the first CES keynote address featuring a healthcare company. In the Abbot presentation, CEO Robert Ford spoke about his vision for "human-powered health." Biowearables "will be like having a window into your body," Ford explained. This technology offers "science that you will be able to access any time so you can understand what your body is telling you and what it needs" (Abbott Health, 2022). Ford concluded the address with a vision for a future with sensors that will empower health consumers, give them more control, more freedom, and less disruption to daily life.

The idea that sensors can provide a window into the body is a common rhetorical phrase used to explain wearable sensor devices. Berg reported encountering this kind of language frequently in his research. Moodmetric and Oura are just two examples of self-tracking devices supposedly designed to promote wellbeing by monitoring body signals that were previously undetected while a person was unaware or sleeping. Though some researchers refute the claim that sleep tracking supports better sleep habits (Zraick et al., 2019), promoters of these devices point to the evidence of millions of satisfied customers. Promotional videos for Moodmetric and Oura support Berg's research concerning the corporate rhetoric that continues to be associated with such products. Further, people are socialized to want to "review data about themselves collected by other actors, such as social media metrics, employee dashboards, educational outcomes, medical records and so on" (Lupton, 2018, p. 1). They are willing to provide the labor and biometric data needed to fuel these platforms. Deborah Lupton (2016; 2018; 2020) has contributed much on the quantified self-movement and the way devices "work to capture and materialise immanent dimensions of human embodiment, creating human–data assemblages" (Lupton, 2018, p. 1).

Corporate rhetoric drives adoption before people can understand its impact. As if they were already legitimate, AI-based technologies are driven by their promised transformative claims to disrupt traditional domains. Clearview AI provides an especially representative example of such a dramatic transformation. Originally a small AI startup, Clearview AI scraped 3 billion public images from Facebook's servers to provide law enforcement agencies with a significant algorithm to identify criminals (Hill, 2020). A vast international collection of people's social media images instantly became a pool of potential suspects. Clearview AI's business model not only exceeded privacy regulations and national laws, it challenged large social media corporations' abilities to control their own platforms. Facebook could no longer control how users would *be used by* their own technology.

Based on a lawsuit in Illinois, in May 2022, Clearview AI agreed in a settlement to stop selling its massive database of images. As discussed by Greg Bensinger (2022):

> The Biometric Information Privacy Act of Illinois sets strict limits on the collection and distribution of personal biometric data, like fingerprints and iris and face scans. The Illinois law is considered among the nation's strongest, because it limits how much data is collected, requires consumers' consent and empowers them to sue the companies directly, a right typically limited to the states themselves.

Therefore, as Clearview AI and other technology companies profit by deploying public images to law enforcement and other private entities, this lawsuit shows that "effective statutes can help bring some of Big Tech's more invasive practices to heel" (Bensinger, 2022).

This is but one example of how augmentation technologies and AI have been emerging over recent decades driven by corporate development, university research, military-industrial complex development, increased biometric data availability, new AI techniques, biological technologies, upgraded computing power, and maturing digital architectures. The emergence of AI has recently accelerated through machine learning algorithms, natural language processing, and predictive models that inform the design of technologies. The *Next Generation of Emerging Global Challenges Horizons 2030* report emphasizes that "the increasing sophistication of physical and cognitive augmentation technologies will unlock new potential for human abilities, health and longevity, potentially raising divisive social, legal, and psychological issues" (Policy Horizons Canada, 2018). Driven largely by such corporate research advancement, technology development, increased data availability, the rise of AI, upgraded computing power, and new architectures, augmentation technologies are emerging largely unchecked and largely not understood by technical communication professionals and scholars/instructors. AI rhetoric contributes to the hype, and policy makers are taking note of it.

This book addresses the expectations emanating from these developments and the proliferation of large companies promising innovations as mainstream phenomena. Elon Musk argues that humans *communicate* and *think* too slowly as he markets Neuralink Corporation's development of brain-implantable neurolace (Ricker, 2016; Orth, 2020). One result is that this kind of corporate-driven efficiency worldview becomes instantiated within the public sphere discourse about technology. It neglects communities, misrepresents cultures, and even harms individuals, whether these technologies are adopted or not (Crawford et al., 2019).

We recognize that global actors constantly alter the way augmentation technologies and AI are adopted, impacting the future and the implications for technical and professional communicators. This book magnifies instances of augmentation technologies and AI in order to appropriately contextualize digital enhancements. We also acknowledge that "algorithmic systems can cause harm when they fail to work as specified—i.e., in error—but may just as well cause real harms when working *exactly* as specified" (Moss et al., 2021).

Moreover, industry reports such as Jim Guszcza, Harvey Lewis, and Peter Evans-Greenwood's (2017) *Deloitte Insights* argue that "humans and computers think better together." In *Writing Futures: Collaborative, Algorithmic, Autonomous* (Duin & Pedersen, 2021), we emphasize the importance of cultivating the ability to write and work alongside augmentation technologies; this includes working with such non-human agents, understanding the impact of algorithms and AI on work and writing, accommodating the unique relationships with autonomous agents, and planning for ongoing disruption. Again, the purpose of this book is to cultivate an even deeper understanding of human augmentation technology

Technical Communication
Future Practice and
Collaboration

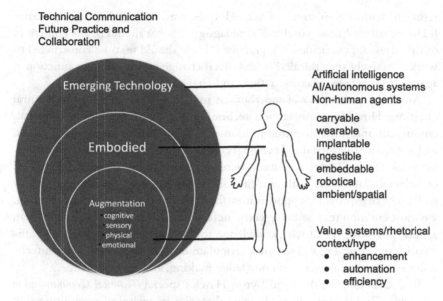

Emerging Technology — Artificial intelligence
AI/Autonomous systems
Non-human agents

Embodied — carryable
wearable
implantable
Ingestible
embeddable
robotical
ambient/spatial

Augmentation
- cognitive
- sensory
- physical
- emotional

Value systems/rhetorical
context/hype
- enhancement
- automation
- efficiency

FIGURE 1.3 Technical communication future practice and collaboration.

and AI, build technical and professional communication capacity to articulate its benefits and risks, and provide direction for future practice and collaboration.

As Figure 1.3 illustrates, technologies influence cognitive, sensory, physical, and emotional states for the purpose of enhancement, efficiency, and automation, the level of which is increasingly impacted by fluctuating value systems, rhetorical context, and corporate hype. Augmentation technologies function within embodied carryable, wearable, implantable, ingestible, embeddable, robotical, and ambient/spatial platforms. Spurred by artificial intelligence, the resulting autonomous systems and non–human agents will increasingly become part of the fabric of our digital lives, the scope of which is detailed in Chapter 2. Given the proliferation of AI and autonomous systems, we contend that TPCs must prepare for promoting human-autonomous teaming surrounding present and emerging augmentation technologies (see Chapter 7).

Rationale for Technical and Professional Communication Engagement with Augmentation Technologies

In 2009, Carolyn Rude's seminal work to map research questions in technical communication provided scholars with solid direction for positioning their work. Rude posited this central question for the field of technical and professional communication: "How do texts (print, digital, multimedia; visual, verbal) and related communication practices mediate knowledge, values, and action in a variety of social and professional contexts?" (p. 176). She positioned research within four areas of related questions surrounding disciplinarity ("How shall we know ourselves?"), pedagogy ("What should be the content of our courses and curriculum?"), practice ("How should texts be constructed to work effectively and ethically?"), and social change ("How do texts function as agents of knowledge making, action, and change?").

Amid the current era of superhuman innovation, we propose this central question: How do augmentation technologies (cognitive, sensory, physical, emotional) and related communication practices mediate knowledge, values, and action in professional and personal contexts? Given the proliferation of augmentation technologies, we must articulate expanded disciplinarity as we find ourselves collaborating with non-human agents; we must develop and deploy pedagogy that provides opportunities for such human-autonomous teaming; we must envision texts as being constructed in collaboration with non-human agents, providing TPCs with knowledge to deploy such work effectively, efficiently, and ethically; and we must articulate how AI and augmentation technologies function as agents of knowledge making, action, and change.

In 2009, Andrew Mara and Byron Hawk's special *Technical Communication Quarterly (TCQ)* issue on posthuman rhetorics in technical communication chronicled the philosophical, scientific, sociological, technological, literary, and cultural discourses surrounding posthumanism as a means to bring posthuman perspectives to bear on critical problems in technical communication: workplace identities, organizational situatedness, human–computer interaction, workplace texts and technologies, pedagogical practices, transitioning past/present/future contexts, and organizational change (p. 1). They wrote:

> As organizations become more complex, technologies more pervasive, and rhetorical intent more diverse, it is no longer tenable to divide the world into human choice and technological or environmental determinism. Professional and technical communication is a field that is perfectly situated to address these concerns. Because it is already predisposed to see the writer in larger organizational contexts, the moment is right to explore technical communication's connections to posthumanism, which works to understand and map these complex rhetorical situations in their broader contexts.
>
> *(p. 3)*

More recently, Kristen Moore and Daniel Richard's (2018) collection, *Posthuman Praxis in Technical Communication*, articulated how posthumanism and praxis together offer direction as we grant increased agency to autonomous, non-human agents. Carl Herndl's (2018) foreword sums it up best:

> The shift to a posthuman praxis in technical communication contributes to making what Haraway calls a more lively and livable world. More

lively because it is populated by agentive things which escape the modern binary between the human and the nonhuman. More livable because accounting for the actions and possibilities of these newly enfranchised things allows us to better manage the interrelations among this reassembled community.

(xiv)

Regarding the future of technical and professional communication, researchers and practitioners also often list emerging trends and associated technologies as a means toward a "more lively and livable world." In his introduction to a special 2005 *Technical Communication* issue on the future of technical communication, Michael Albers argued that:

the future of the field will be technology laden. Technology permeates everything a practicing technical communicator does. How we react to changes in that technology on both the individual and organizational level will have a dramatic impact on the development of the profession.

(p. 271)

Chris Lam (2021), in his recent study of Society of Technical Communication members, networks, trends, and themes from 2016 to 2019 through examination of 75,333 Tweets using the hashtag #TechComm, states that:

Technical communication is a field that continually evolves because of its connection to technology—as technology evolves, so too does communication with and about technology.... [Its] competencies, skillsets, and areas of expertise, continues to evolve as related fields and sub-disciplines emerge like content strategy and user experience (UX).

(p. 6)

Lam articulates dominant themes as including the Adobe TC suite and other authoring tools, the Society of Technical Communication (STC) summit, academic tech comm, professional development, DITA and lightweight DITA, job opportunities in tech comm, and software documentation. Not surprisingly, Lam identifies only limited overlap between academic and practitioner communities, with the word "rhetoric" "tweeted almost exclusively by academics" (p. 17). This aligns with related studies of academic and practitioner journals and magazines (Boettger & Friess, 2016; Andersen & Hackos, 2018). The strongest focus on technology appears within "the practitioner orbit," and Lam identifies XML and DITA instruction as being foundational for the field (note Duin and Tham's 2018 development of such curricula).

Lam also identifies Scott Abel, known as The Content Wrangler, as the practitioner with the largest lively audience: 20,600 followers. In 2019, Abel hosted a four-week Society for Technical Communication Roundtable discussion on the future of technical communication in which participants examined trends, challenges, methods, standards, tools, and technologies in use across technical communication departments that spanned over 600 professionals working in 16 countries. Respondents were "super excited" about the future impact of advanced technologies on the way they live and work, being most excited about chatbots, with 50% of teams planning to launch a bot by 2020, along with voice assistants and AI. They simultaneously anticipated and feared AR, with one director of technical information for a medical device manufacturer stating, "Augmented reality will radically transform how technical and scientific information is communicated, bringing technical documentation content directly to—and integrating it with—the point-of-use."

In 2020, Abel invited Rob Gillespie, information architect from the UK, to share his thoughts in a webinar about the future of technical communication (Zoomin, 2020). Gillespie speaks of the unique role of the technical communicator as a "value generating entity" and the strategic role as technical communication becomes more integrated with and integral to the business. He emphasizes the importance of working closely and concisely with and for machines; being careful about what to input into the machine; and expanding the use of tools to "make us better." He stresses that the "old tools and cubbyholes" are gone and that technical communicators should plan for working with robots and AI. Gillespie promotes three steps for preparing for the future: (1) to be proactive about collaboration, making sure that processes are well defined and efficient; (2) to embrace robotic automation and AI, stating that "they are our friends, [we] must work with them and understand how they work"; and (3) to automate everything that can be automated, emphasizing that technical communicators need to collaborate to move "laborious tasks" to machines.

As a result, while increasing numbers of programs are working to integrate XML and DITA instruction, we see it as imperative for technical communication researchers, instructors, and practitioners to understand how augmentation technologies are modeled to be integrated within our personal and professional lives and proposed to enhance and transform our thinking, sensing, and feeling. It is critical to address the evolving ethical and rhetorical dimensions and dilemmas surrounding them. Therefore, we construct this book as a guide for preparing technical communication practitioners and instructors to cultivate understanding of augmentation technologies, build digital and AI literacy, and design ethical futures. Moreover, we work to provide direction for reframing understanding of audience, usability, and both academic and practitioner roles in technical and professional communication amid the rapid emergence of augmentation technologies.

However, one of the greatest challenges facing technical and professional communication scholars and instructors continues to be a reticence to prepare for such advance of major technological transformations. Citing this urgent need, in *Writing Futures: Collaborative, Algorithmic, Autonomous* (Duin & Pedersen, 2021) we introduced a *Writing Futures* Framework for scholars and instructors to investigate and plan for social, digital literacy, and civic implications of collaborative, algorithmic, and autonomous writing futures. Our goal was to provide readers with opportunities to begin to understand and write alongside non-human agents, examine the impact of algorithms and AI on writing, and accommodate the unique relationships emerging with autonomous agents. *Augmentation Technologies and Artificial Intelligence in Technical Communication: Designing Ethical Futures* provides greater depth for understanding augmentation technologies and, amid enhancement of professional capabilities, a strong focus on building technical and professional communication ability, and strategic knowledge to articulate its benefits, risks, and relevance.

We understand that readers across our broad technical and professional communication field may well not hold a unified view of professionalization for work with augmentation technologies and artificial intelligence. Saul Carliner (2012) explored such tension, proposing a spectrum from *formal professionalism* "rooted in a worldview that values expertise" to *quasi professionalization* "in which individuals participate in the activities of the occupational infrastructure but without the expectation of exclusive rights to perform the work" to *contra-professionalization*, in which professionals offer services "outside of parts of the entire infrastructure, sometimes circumventing it completely" (p. 49). At this current time of technological transformation, throughout this book we offer direction for building professionalization, and we agree with Carliner in that such current lack of consensus on the TPC infrastructure for understanding, design, and adoption of augmentation technologies and AI, may well result in "a competitive environment for certain types of high-value assignments" (p. 49) as these technologies emerge. The last decade indeed has further accentuated the need for TPCs to evolve professional understanding of audience, user experience, content management, collaboration, and design.

Moreover, this book involves multiple contours of posthumanism as we increasingly participate in complex assemblages with non-human agents through distributed services and digital platforms. George Hayhoe and Pam Estes Brewer (2021), in *A Research Primer for Technical Communication*, discuss the various goals that direct research, citing Thomas Reeves' (1998) six categories of research: theoretical, empirical, interpretivist, postmodern, developmental, and evaluative. Of these we focus most on postmodern research, "examining the assumptions that underlie applications of technology or technical communication with the ultimate goal of revealing hidden agendas and empowering disenfranchised groups" (Hayhoe & Brewer, p. 7).

Table 1.1 provides an initial summary of how core functions of technical communication—audience, usability, content development, collaboration, and design—transform alongside augmentation technologies. These core functions evolve as AI and use of non-human agents become increasingly integrated

TABLE 1.1 Current and future core functions of technical and professional communication

Core function	Current	Future
Audience addressed, invoked, involved	Audience immersed	Audience [agency] augmented
User/human experience (UX)	Augmentation technology as UX team support, interdependent team member	Non-human agent as independent team member, digital employee
Content development	Topic-based content management systems	Machine learning and explainable AI
Collaboration	Participatory design Agile development	Human-autonomy teaming
Dialogic design	Social justice design ethics Surveillance culture	AI ethics Algorithmic impact assessment

throughout TPC work. With such integration, we need to increase focus on the audience as immersed (Tham, 2018) and now augmented through increased awareness as a result of augmentation technology use. User experience and usability testing evolve to include augmentation technology support and increased use of non-human agents as team members. Content management systems and associated standards (e.g., DITA) will include increased use of machine learning and explainable AI; and collaboration will increasingly depend on human-autonomy teaming.

Throughout this evolutionary change, TPCs will continue to be tasked with understanding and documenting augmentation technologies to make them understandable and usable by professionals and the public. Too often TPCs are provided with a combination of marketing information and/or technical specifications from subject matter experts, with neither set of documents correctly or adequately representing what users need.

Related disciplines are studying this evolutionary change. Shyam Sundar and Eun-Ju Lee (2022) edited a recent issue of *Human Communication Research* in which contributing authors explored the role of AI in communication along conceptual dimensions in human–computer interaction (HCI) and computer-mediated communication (CMC) research, describing how AI can fulfill analogous roles of communicator or mediator. They ask, "How well can AI replace a human in serving as a communicator?" Here we explore augmentation technologies and AI, with focus on human–AI collaboration. Throughout this book, we describe how technical and professional communication roles will change as a result of increased use of augmentation technologies and AI. We agree with Sundar and Lee in that AI technology "has profoundly affected virtually all areas of our lives over the past decade" (p. 379). In their case, the traditional distinctions between HCI and CMC now must be modified, given AI involvement. In our case, traditional approaches to audience, purpose, and intended effects now must be modified. Their special issue was conceived:

> to address the questions of how such wide-spread integration of AI in communication might alter the very essence of human communication and force communication scholars to revisit widely accepted assumptions and knowledge about how we humans communicate, with what consequences.
>
> *(p. 381)*

In this book, we provide rationale and direction for positioning TPCs to understand augmentation technologies and AI as a means to design ethical futures across this work.

Moreover, the Fabric of Digital Life, or "Fabric" (https://fabricofdigitallife. com), detailed in Chapter 2, provides a means to access and understand the metadata behind an emerging technology, resulting in deeper knowledge of modes of technology invention over time and how technologies move from the seeming fringe to mainstream use. We advocate for greater understanding of metadata as a means for TPCs to better understand augmentation technologies, resulting in more informed persona development and information design. Examining metadata allows us to uncover sociotechnical tradeoffs in the technologies (Iliadis & Pedersen, 2018). The burgeoning augmentation technologies market masks complex sociotechnical tradeoffs that TPCs must understand as part of their work to develop usable and ethical content. For example, a business will frame an augmentation technology as a business solution without conveying critical information about the concessions that users must make in terms of giving up their data in exchange for access and use. TPCs can examine Fabric's rich metadata fields to search for terms and technologies as a means to better understand augmentation technologies and the potential tradeoffs related to usability.

Rhetorical scholar K. J. Rawson (2018) explored how such curated archival description results in greater bureaucratic and epistemological understanding, providing an extended consideration of the *description* archival process. Rawson quotes MacNeil (2009, p. 90) stating that archival description:

> involves telling a story about records, which both 'changes the meaning of the records' and 'determine[s] how they will be used and remade in the future.' Such transformations evidence the rhetorical power of archival description…. It is created to suit particular audiences and needs; and it can have tremendous influence over the reception and use of the materials it describes.
>
> *(pp. 328–329)*

Thus, the archival descriptions in Fabric, i.e. the collection descriptions and the metadata for each digital artifact, provide tremendous understanding of the reception and use of augmentation technologies. TPCs are well positioned to understand metadata as a means to unveil and address the intricacies of augmentation technologies. In Chapter 2 we illustrate how we use Fabric of Digital Life metadata to help define augmentation technologies, trace them over time, and provide a means to prepare for the present and the future.

Augmentation Technologies—Designing Ethical Futures Framework

The core of our human experience, identity, and professional and personal practices is affected by the momentum of augmentation technologies and AI that enhance human capabilities or productivity by adding to the body (or ambient environment around the body) cognitive, physical, sensory, and/or emotional enhancements. Emerging wearable devices give people with sight loss the ability to see their environment; earbuds translate conversations in real time; nanobots deployed into the body deliver drugs to target and attack disease; and neural implants transmit brain activity to allow humans to control machines using only thoughts.

This book brings theoretical, empirical, pedagogical, critical, and ethical attention to the development of these sophisticated, emergent, and embodied augmentation technologies slated to promote enhanced futures—i.e., to improve lives, literacy, cultures, arts, economies, and social and professional contexts. To date, the study of the emergence of augmentation technologies largely has neglected to adopt such an ethical, human-centric approach, leaving citizens, instructors, practitioners, and governments to deal with the consequences. Moreover, while technical and professional communication has begun to come to terms with socio-ethical problems and inequities (Walton et al., 2019), augmentation technology and AI design, adoption, and adaptation have not followed suit, nor have they properly dealt with grand challenges including algorithmic bias, lack of diversity in development teams, and misuse of augmentation technologies and AI in the field.

Figure 1.4 provides an overview of the major sections and associated themes in this book. As shown in the center of this visual, the purpose of this work is to reframe professional practice and pedagogy to promote digital and AI literacy surrounding the ethical design, adoption, and adaptation of augmentation technologies. The three larger shaded boxes indicate the major sections in the book; the lighter shaded boxes indicate key themes presented in each section:

- Understand (rhetorics of) augmentation technologies
  - Introduces the Ethical Futures Framework, defines and articulates the dimensions of augmentation technologies, focuses on the development

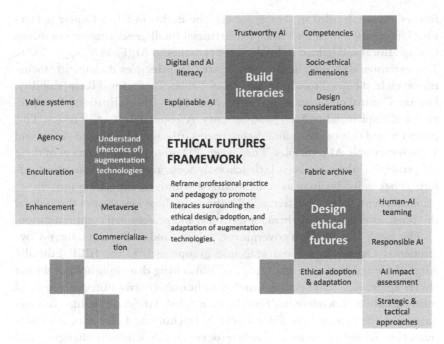

FIGURE 1.4 Ethical futures framework.

of non-human agents in industry as a critical factor in the rise of augmentation technologies, and provides critical attention to hyped assumptions and recent claims concerning the Metaverse and AR.

- Build literacies
  - Provides detailed definition of digital and AI literacy amid the emergence of augmentation technologies, with focus on examining, exploring, and participating in the development of digital and AI literacy skills needed to address algorithmic mining and bias, racial discrimination, digital divides, unethical AI practices, misinformation, and other socio-ethical harms to humans.
- Design ethical futures
  - Provides direction for cultivating digital literacy and AI literacy skills needed to assess integration and participation with augmentation technologies, chronicles the rapid increase in autonomous agents and digital employees, positions TPCs for intervening throughout the design, adoption, and adaptation of augmentation technologies, and shares strategic and tactical approaches for designing ethical augmentation technology and AI futures.

Only by understanding and embedding ethical principles throughout augmentation technology design and AI can we foster human-centered, humane futures. As we detailed in *Writing Futures*, the Berkman Klein Center at Harvard University started the Principled Artificial Intelligence Project as a means to map ethical and human rights–based approaches to AI (Fjeld & Nagy, 2020). The extensive data visualization synthesizes 36 principles documents, focusing on eight themes: Promotion of Human Values, Professional Responsibility, Human Control of Technology, Fairness and Non-Discrimination, Transparency and Explainability, Safety and Security, Accountability, and Privacy. This project found that organizations do not necessarily agree on how to define and apply responsible AI principles. They write, "Despite the proliferation of these 'AI principles,' there has been little scholarly focus on understanding these efforts either individually or as contextualized within an expanding universe of principles with discernible trends." We draw on these key themes throughout this book, contextualizing them for technical and professional communication.

This book also draws on government, private, and civil sectors, Inter-Governmental Organizations, and academic groups such as the IEEE Ethically Aligned Design initiative, analyzing and illustrating that rightsholders do not always agree; e.g., some entities emphasize human control of technology and others ignore it. Therefore, we find the Principled Artificial Intelligence visualization to be extremely useful in research, teaching, and outreach, as it illustrates how AI values can be seen as heterogeneous, dynamically changing, and always contextualized. Figure 1.5 from the top of this visualization includes the key themes and reports from civil society and government organizations, themes that we discuss in detail in Chapter 5.

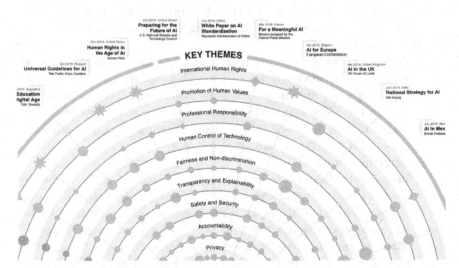

FIGURE 1.5   A section from the principled artificial intelligence visualization, Berkman Klein Center at Harvard University. Creative Commons Attribution 3.0 Unported license. https://creativecommons.org/licenses/by/3.0/.

Indeed, researchers in AI ethics, robot ethics, and philosophy are actively delineating value-based frameworks for embedding synthetic emotions in augmentation technologies. Throughout this book, we also invite you to use the key IEEE resource *Ethically Aligned Design: A Vision for Prioritizing Human Well-being with Artificial Intelligence and Autonomous Systems, First Edition.* As the report states, its purpose is:

> to establish frameworks to guide and inform dialogue and debate around the non-technical implications of these technologies, in particular related to ethical aspects. We understand 'ethical' to go beyond moral constructs and include social fairness, environmental sustainability, and our desire for self-determination.
>
> *(The IEEE Global Initiative on Ethics of Autonomous and Intelligent Systems, 2019, p. 3).*

Sections and Chapters

The sections of this book align with the three sections of the Ethical Futures Framework. Section 1 concentrates primarily on understanding the rhetorics of augmentation technologies. In this chapter we have provided an initial definition of augmentation technologies along with prominent scholarly directions that inform its development. We have included rationale for positioning technical and professional communication researchers, instructors, and practitioners to understand and articulate future changes to core technical communication functions of audience, user experience, content development, collaboration, and design as impacted by augmentation technologies and AI.

Chapter 2 identifies a changing set of value systems that constitute augmentation technologies as social phenomena, namely, beliefs surrounding enhancement, automation, and efficiency. It articulates and examines cognitive, sensory, emotional, and physical enhancements as a range of subcategories of augmentation technologies and AI. It illustrates how the Fabric of Digital Life (https://fabricofdigitallife.com) archive's rich metadata fields provide a means to examine augmentation technologies and the complex sociotechnical tradeoffs that technical and professional communicators must navigate as they work to develop usable content and direction.

Chapter 3 provides critical attention to the hyped assumption that sophisticated, emergent, and embodied augmentation technologies will improve lives, literacy, cultures, arts, economies, and social contexts. It begins with discussion of the problem of ambiguity with AI terminology, providing a description of the WIPO Categorization of AI Technologies Scheme to assist with it. This chapter then draws on media and communication studies to explore concepts such as agents, agency, power, and agentive relationships between humans and robots. The chapter focuses on the development of non-human agents in industry as a critical factor in the rise of augmentation technologies. It looks at how marketing communication enculturates future users to adopt and adapt to the technology. Scholars are charting the significant ways that people are drawn further into commercial digital landscapes, such as the Metaverse concept, in post-internet society. It concludes by examining recent claims concerning Metaverse and AR through the use of an electronic soft contact lens platform.

Section 2 concentrates primarily on pedagogy and guiding principles for building digital and AI literacy. Chapter 4 defines digital and AI literacy for current and future work with augmentation technologies. The exponential increase in AI techniques, functional applications, and use across application fields demands critical attention to digital literacy, data literacy, AI literacy, AI explainability, and trustworthy AI. This chapter uses Long and Magerko's (2020) definition of AI literacy and conceptual framework for determining AI competencies and design considerations as a means for TPC scholars, instructors, students, and practitioners to examine and develop digital and AI literacies. It focuses on explainable AI (XAI) as a means to help humans better understand how AI works and makes particular decisions. It then discusses the Assessment List for Trustworthy AI (ALTAI) found at the European AI Alliance portal (Welcome to the ALTAI portal!, 2020), a set of strategic questions for use as an initial approach to evaluating trustworthy AI for the purpose of minimizing risks while maximizing the benefits of AI for human users. The chapter concludes with discussion of how technical and professional communication roles are evolving as a result of augmentation technologies and AI.

Chapter 5 addresses potential socio-ethical consequences of augmentation technologies and AI. It provides clarity on biometric data and its relevance to augmentation technologies in combination with artificial intelligence. *Principled Artificial Intelligence: Mapping Consensus in Ethical and Rights-Based Approaches to Principles for AI* (Fjeld et al., 2020) provides the ethical focus. The chapter uses work published by the AI Now Institute, which monitors and reports on human rights issues through several published reports. These include focus on algorithmic mining and bias, racial and gender discrimination, ableism, digital divides, unethical AI practices, misinformation, and other socio-ethical harms to humans. We continue to build rationale for understanding digital and AI literacy as a means to avoid socio-ethical harms to humans that occur when digital and AI literacy is absent.

Section 3 concentrates primarily on practitioner direction, beginning in Chapter 6 with discussion of pedagogical direction for cultivating digital literacy and AI literacy skills needed to assess integration and participation with augmentation technologies. We draw on multiple studies on building digital literacy (Burnham & Tham, 2021; Davis et al., 2021; Duin et al., 2021; Tham et al., 2021) to provide direction for cultivating digital and AI literacy through examining and curating augmentation technologies. We describe student and instructor perception of digital and AI literacy through studies of learner engagement with the Fabric of Digital Life archive/artifacts along with theoretical and pragmatic model development for instructor-scholars to design and integrate digital and AI literacy development that caters to student learning needs as well as their professional futures and workforce preparedness. The chapter provides suggestions for integrating study of augmentation technologies in technical communication across undergraduate and graduate curricula, concluding with discussion of learner capabilities for a world with augmentation technologies and AI.

Chapter 7 begins with discussion of guidelines for professional practice surrounding human-AI interaction that includes a TPC guide to human-AI interaction based on key studies in human-centered computing, science and engineering, and technical communication. Given the rapid increase in autonomous agents and human-AI interaction, the chapter's main focus is on seeing tools as autonomous agents or digital employees. After discussion of chatbots and conversational design, the chapter chronicles the development of digital employees across six corporations. Based on AI functional technologies, these digital employees are proposed to enhance meaningful, empathetic connections to the digital world, resulting in potentially greater symbiotic relationships with users. TPCs must be positioned to intervene throughout the design, adoption, and adaptation of augmentation technologies. The audience is now an "augmented" audience; autonomous agents now function as independent team members; and content is produced through machine learning.

Chapter 8 focuses on designing ethical futures by way of strategic and tactical approaches to governance, regulation, and standardization of augmentation technologies and AI. Influenced by philosopher Michel de Certeau's (1984) distinction between strategies and tactics, the chapter concentrates on TPC capabilities to enact changes in their field toward ethical ends. The chapter begins by noting exemplary organizations working to promote ethical design of augmentation technologies, whose collective policy guidelines work to address risk. It then focuses on risk communication and awareness of ethical algorithmic impact assessment tools and processes to help guide design of and collaboration with augmentation technologies, including digital employees. Strategic approaches include work underway in the European Union and proposed in the US. Tactical approaches discussed include coalition and standards building, including Metaverse professional standards bodies currently forming to help build foundations for open standards among corporations, and a journey mindset. The chapter concludes with an invitation to collaborate on research underway as part of the Digital Life Institute at https://www.digitallife.org/.

To conclude, professionals across multiple industries, including at IBM (IBM, 2022), Microsoft (Spencer, 2019), and Samsung (Samsung Advanced Institute of Technology, n.d.), welcome multi-disciplinary direction for navigating the evolving augmented technology landscape. For example, Samsung is proposing an exoskeleton technology concept for fitness and entertainment to be used in the home. Samsung's Innovation Campus (SIC) offers AI development education and teaching materials to youth around the world, and the annual CES trade show, "the world's most influential technology event," promotes and celebrates augmentation technologies far in advance of their actual deployment. While futuristic model cars and future homes are often on display, so too are human augmentation technologies. The January 2020 CES featured a session called "The Next Era: Superhuman Innovation" asking "What if we could tap into superhuman powers to be better at school, excel at sports, succeed in business and ultimately live a longer and fuller life? Who wouldn't want that competitive edge?" (Duffey, 2019). And the January 2022 CES featured sessions on biowearables of the future, smart contact lenses, and access to the Metaverse via Vuzix Shield AR glasses.

Moreover, corporations seek direction as they are tasked with requirements to develop algorithmic impact assessment tools and processes to ensure transparency and accountability for decision-making systems (Reisman et al., 2018). The US Algorithmic Accountability Act of 2022, as summarized at https://www.wyden. senate.gov, "requires companies to assess the impacts of the automated systems they use and sell, creates new transparency about when and how automated systems are used, and empowers consumers to make informed choices about the automation of critical decisions." Working in tandem with usability experts, with this book's direction, technical and professional communication scholars and practitioners will be positioned to meet increased expectations for the ethical design, adoption, and adaptation of augmentation technologies and AI.

References

Abel, S. (2019). *The future of tech comm.* Society for Technical Communication roundtable. https://www.stc.org/roundtable/future-of-tech-comm/#1553631912343-27d700db-87f1

Abbott Health. (2022, January 20). *Abbott health—Next biowearables of the future* [Video]. YouTube. https://www.youtube.com/watch?v=oMCPdaCbO2Q&t=183s

Albers, M. J. (2005). The future of technical communication: Introduction to this special issue. *Technical Communication, 52*(3), 267–272.

Algorithmic accountability act of 2022 [Summary brief]. (2022). Ron Wyden U.S. Senator for Oregon. https://www.wyden.senate.gov/download/one-pager-bill-summary-of-the-algorithmic-accountability-act-of-2022

Andersen, R., & Hackos, J. (2018). *Increasing the value and accessibility of academic research: Perspectives from industry* [Conference session]. SIGDOC '18, Association of Computing Machinery, Milwaukee, WI, USA.

Andrejevic, M., & Burdon, M. (2014, April 11). Defining the sensor society. *Television and new media.* University of Queensland TC Beirne School of Law Research Paper. No. 14–21. https://ssrn.com/abstract=2423118

Barad, K. (2003). Posthumanist performativity: Toward an understanding of how matter comes to matter. *Signs, 28*(3), 801–831. https://doi.org/10.1086/345321

Bensinger, G. (2022, May 30). How Illinois is winning in the fight against big tech. *New York Times.* https://www.nytimes.com/2022/05/30/opinion/illinois-biometric-data-privacy.html

Berg, M. (2017). Making sense with sensors: Self-tracking and the temporalities of wellbeing. *Digital Health, 3,* 1–11. https://doi.org/10.1177%2F2055207617699767

Boettger, R. K., & Friess, E. (2016). Academics are from Mars, practitioners are from Venus: Analyzing content alignment within technical communication forums. *Technical Communication, 63*(4), 314–327.

Braidotti, R. (2013). *The posthuman.* Polity.

Brooker, C. (Creator). (2011–2019). *Black Mirror* [TV series]. Zeppotron (2011–2013), House of Tomorrow (2014–2019).

Burdon, M. (TEDxUQ) (2015). *The collected or the collective? Privacy in a sensor society.* YouTube. https://www.youtube.com/watch?v=KtO4_YD5ThM

Burnham, K., & Tham, J. (2021). Developing digital literacy through multi-institution collaboration and technology partnership: Analysis of assignments, student responses, and instructor reflections. *Programmatic Perspectives, 12*(2), 59–100.

Carliner, S. (2012). The three approaches to professionalization in technical communication. *Technical Communication, 59*(1), 49–65.

Clark, A. (1997). *Being there: Putting brain, body, and world together again.* MIT Press.

Clark, A. (2003). *Natural born cyborgs: Minds, technologies, and the future of human intelligence.* Oxford University Press.

Crawford, K., Dobbe, R., Dryer T., Fried, G., Green, B., Kaziunas, E., Kak, A., Mathur, V., McElroy, E., Sánchez, A. N., Raji, D., Rankin, J. L., Richardson, R., Schultz, J., West, S. M., & Whittaker, M. (2019). *AI NOW 2019 Report.* AI NOW Institute. https://ainowinstitute.org/AI_Now_2019_Report.html

Davis, K., Hocutt, D., Stambler, D., Veeramoothoo, C., Tham, J., Duin, A. H., Ranade, N., Misak, J., & Pedersen, I. (2021, August 25). Fostering student digital literacy through the Fabric of Digital Life. *Journal of Interactive Technology and Pedagogy.* https://jitp.commons.gc.cuny.edu/fostering-student-digital-literacy-through-the-fabric-of-digital-life/

Dial, M. (2018). *Heartificial empathy: Putting heart into business and artificial intelligence.* DigitalProof Press.

de Certeau, M. (1984). *The practice of everyday life,* trans. S. Rendall. University of California Press.

Duffey, C. (2019, December 20). *The next era: Superhuman innovation.* CES. https://www.ces.tech/Articles/2019/The-Next-Era-Superhuman-Innovation.aspx

Duin, A. H., & Pedersen, I. (2021). *Writing futures: Collaborative, algorithmic, autonomous.* Springer. https://www.springer.com/gp/book/9783030709273

Duin, A. H., & Pedersen, I. (2023). *Augmentation technologies and AI—An ethical design futures framework* [multimedia collection]. Fabric of Digital Life. https://fabricofdigitallife.com/Browse/objects/facets/collection:69

Duin, A. H., Pedersen, I., & Tham, J. (2021). Building digital literacy through exploration and curation of emerging technologies: A networked learning collaborative. In N. B. Dohn, J. J. Hansen, S. B. Hansen, T. Ryberg, & M. de Laat (Eds.), *Conceptualizing and innovating education and work with networked learning* (pp. 93–114). *Research in Networked Learning.* Springer. https://doi.org/10.1007/978-3-030-85241-2_6

Duin, A.H., & Tham, J. (2018). Cultivating code literacy: A case study of course re-design through advisory board engagement. *Communication Design Quarterly, 6*(3), 44–58.

Emoshape Inc. (2018, May 3). *Emotion chip EPU for real-time emotion synthesis and reasoning* [Video]. YouTube. https://www.youtube.com/watch?v=OUYRBrPG9ZA

The entire history of you. (2021). In *Wikipedia*. Retrieved from https://en.wikipedia.org/w/index.php?title=The_Entire_History_of_You&oldid=1060082311

Fjeld, J., & Nagy, A. (2020). *Principled artificial intelligence: Mapping consensus in ethical and rights-based approaches to principles for AI*. Berkman Klein Center for Internet & Society at Harvard University. https://cyber.harvard.edu/publication/2020/principled-ai

Fjeld, J., Achten, N., Hilligoss, H., Nagy, A., & Srikumar, M. (2020). *Principled artificial intelligence: Mapping consensus in ethical and rights-based approaches to principles for AI*. Berkman Klein Center Research Publication. https://ssrn.com/abstract=3518482

Forbes Technology Council Expert Panel. (2020, May 14). *12 tech leaders discuss the most intriguing applications of human augmentation*. Forbes Technology Council. https://www.forbes.com/sites/forbestechcouncil/2020/05/14/12-tech-leaders-discuss-the-most-intriguing-applications-of-human-augmentation/

Gartner Inc. (2020, August 18). *Gartner identifies five emerging trends that will drive technology innovation for the next decade* [Press release]. https://www.gartner.com/en/newsroom/press-releases/2020-08-18-gartner-identifies-five-emerging-trends-that-will-drive-technology-innovation-for-the-next-decade

Guibeault, D., & Finkelstein, J. (2019). Human-bot ecologies. In Z. Papacharissi (Ed.), *A networked self and human augmentics, artificial intelligence, sentience* (pp. 152–186). Taylor & Francis.

Guszcza, J., Lewis, H., & Evans-Greenwood, P. (2017, January 23). *Cognitive collaboration: Why humans and computers think better together*. Deloitte Insights. https://www2.deloitte.com/us/en/insights/deloitte-review/issue-20/augmented-intelligence-human-computer-collaboration.html

Hansell, G.R., & Grassie, W. (2011). *H+/-: Transhumanism and its critics*. Metanexis.

Haraway, D. (1985). A manifesto for cyborgs: Science, technology, and socialist feminism in the 1980s. *Socialist Review, 80*, 65–108. https://doi.org/10.1080/08164649.1987.9961538

Harbisson, N. (n.d.). *Neil Harbisson: Cyborg and transpecies activist*. Swiss Re. https://www.swissre.com/profile/Neil_Harbisson/ep_d26e7f

Hartzog, W. (2015). Unfair and deceptive robots. *Maryland Law Review, 74*(4), 785–829.

Hayhoe, G. F., & Brewer, P. E. (2021). *A research primer for technical communication: Methods, exemplars, and analyses* (2nd edition). Routledge.

Hayles, N. K. (1999). *How we became posthuman: Virtual bodies in cybernetics, literature, and informatics*. University of Chicago Press.

Hayles, N. K. (2002). Flesh and metal: Reconfiguring the mindbody in virtual environments. *Configurations, 10*, 297–320.

Herndl, C. (2018). Foreword. In K. Moore, & D. Richard (Eds.), *Posthuman praxis in technical communication* (pp. xi–xiv). Taylor & Francis.

Hill, K. (2020, January 18). The secretive company that might end privacy as we know it. *New York Times*. https://www.nytimes.com/2020/01/18/technology/clearview-privacy-facial-recognition.html

Hurwitz, J., Morris, H., Sidner, C., & Kirsch, D. (2020). *Augmented intelligence: The business power of human-machine collaboration*. CRC Press: Taylor & Francis Group.

The IEEE Global Initiative on Ethics of Autonomous and Intelligent Systems. (2019). *Ethically aligned design: A vision for prioritizing human well-being with autonomous and intelligent systems, version II.* https://ethicsinaction.ieee.org/wp-content/uploads/ead1e.pdf

IBM. (2022). *Forward thinking: Experts reveal what's next for AI.* https://www.ibm.com/watson/advantage-reports/future-of-artificial-intelligence.html

Iliadis, A., & Pedersen, I. (2018). The fabric of digital life: Uncovering sociotechnical tradeoffs in embodied computing through metadata. *Journal of Information Communication and Ethics in Society, 16*(3), 311–327. http://dx.doi.org/10.1108/JICES-03-2018-0022

Kay, G. (2021, April 12). *Elon Musk's Neuralink could transition from implanting chips in monkeys to humans within the year.* Business Insider. https://www.businessinsider.com/elon-musk-predicts-neuralink-chip-human-brain-trials-possible-2021-2021-2

Kenyon, R. V., & Leigh, J. (2011). *Human augmentics: Augmenting human evolution.* 33rd Annual International Conference of the IEEE EMBS, Boston, MA, 2011, 6758–6761.

Kundan, N. N., Patil, A., & Kumar, V. (2021). *Sensor market.* Allied Market Research. https://www.alliedmarketresearch.com/sensor-market

Kurzweil, R., & Grossman, T. (2009). Fantastic voyage: live long enough to live forever. The science behind radical life extension questions and answers. *Studies in Health Technology and Informatics, 149,* 187–194.

Lam, C. (2021). Hashtag #TechComm: An overview of members, networks, and themes from 2016–2019. *Technical Communication, 68*(2), 5–21.

Levy, S. (2021). A new company pursues total recall—Starting with Zoom. *Wired.* https://www.wired.com/story/plaintext-new-company-total-recall-zoom/

Long, D., & Magerko, B. (2020). *What is AI literacy? Competencies and design considerations.* CHI '20, April 25–30, 2020, Honolulu, HI, USA. https://doi.org/10.1145/3313831.3376727

Lupton, D. (2016). *The quantified self.* John Wiley & Sons.

Lupton, D. (2018). How do data come to matter? Living and becoming with personal data. *Big Data & Society, 5*(2), 1–11.

Lupton, D. (2020). Wearable devices: Sociotechnical imaginaries and agential capacities. In I. Pedersen & A. Iliadis (Eds.), *Embodied computing: Wearables, implantables, embeddables, ingestibles* (pp. 49–69). MIT Press.

MacNeil, H. (2009). Trusting description: Authenticity, accountability, and archival description standards. *Journal of Archival Organization, 7,* 89–107.

Mara, A., & Hawk, B. (2009). Posthuman rhetorics and technical communication. *Technical Communication Quarterly, 19*(1), 1–10.

McKee, H. A., & Porter, J. E. (2017). *Professional communication and network interaction: A rhetorical and ethical approach.* Taylor & Francis.

McKee, H. A., & Porter, J. E. (2020). Ethics for AI writing: The importance of rhetorical context. In *Proceedings of the AAAI/ACM Conference on AI, Ethics, and Society* (AIES '20), February 7–9, 2020, New York, NY, USA, pp. 110–116.

McKee, H.A., & Porter, J.E. (2022). Team roles & rhetorical intelligence in human-machine writing. In *Proceedings of the 2022 IEEE International Professional Communication Conference (ProComm),* July 17–20, 2022, Limerick, Ireland, pp. 384–391.

Michael, K., Michael, M. G., Perakslis, C., & Abbas, R. (2020). Uberveillance and the rise of last-mile implantables: Past, present, and future. In I. Pedersen & A. Iliadis (Eds.), *Embodied computing: Wearables, implantables, embeddables, ingestibles* (pp. 97–130). MIT Press.

MIT Sloan Management Review, in collaboration with Deloitte. (2021). *The workforce ecosystem perspective*. MIT Sloan Management Review. https://sloanreview.mit.edu/interactive-the-workforce-ecosystem-perspective/

Moon Ribas. (n.d.). Propela. https://www.propela.co.uk/moon-ribas

Moore, K., & Richard, D. (Eds.). (2018). *Posthuman praxis in technical communication*. Taylor & Francis.

Moss, E., Watkins, E. A., Singh, R., Elish, M. C., & Metcalf, J. (2021) *Assembling accountability: Algorithmic impact assessment for the public interest*. Data and Society Report. https://datasociety.net/library/assembling-accountability-algorithmic-impact-assessment-for-the-public-interest/

Novak, J., Archer, J., Mateevitsi, V., & Jones, S. (2016). Communication, machines & human augmentics. *Communication +1, 5*(1), Article 8. https://scholarworks.umass.edu/cpo/vol5/iss1/8

Orth, M. (2020). TechnoSupremacy and the final frontier: Other minds. In I. Pedersen & A. Iliadis (Eds.), *Embodied computing: Wearables, implantables, embeddables, ingestibles* (pp. 211–235). MIT Press.

Papacharissi, Z. (Ed.). (2019). *A networked self and human augmentics, artificial intelligence, sentience*. Taylor & Francis.

Pedersen, I., & Iliadis, A. (Eds.). (2020). *Embodied computing: Wearables, implantables, embeddables, ingestibles*. MIT Press.

Pedersen, I., & Mirrlees, T. (2017). Exoskeletons, transhumanism, and culture: Performing superhuman feats. *IEEE Technology & Society Magazine, 36*(1), 37–45.

Pfister, D. S., & Yang, M. (2018, August 14). Five theses on technoliberalism and the networked public sphere. *Communication and the Public*. https://doi.org/10.1177/2057047318794963

Policy Horizons (2018, October 19). *The next generation of emerging global challenges horizons 2030*. https://horizons.gc.ca/en/2018/10/19/the-next-generation-of-emerging-global-challenges/

Rawson, K. J. (2018). The rhetorical power of archival description: Classifying images of gender transgression. *Rhetoric Society Quarterly, 48*(4), 327–351.

Reeves, T. (1998). The scope and standards of the. *Journal of Interactive Learning Research*. https://www.aace.org/pubs/jilr/scope/

Reisman, D., Schultz, J. Crawford, K., & Whittaker, M. (2018, April). *Algorithmic impact assessments: A practical framework for public agency accountability*. AI Now. https://www.ftc.gov/system/files/documents/public_comments/2018/08/ftc-2018-0048-d-0044-155168.pdf

Ricker, T. (2016). Elon Musk: We're already cyborgs. *The Verge*. https://www.theverge.com/2016/6/2/11837854/neural-lace-cyborgs-elon-musk

Rude, C. (2009). Mapping the research questions in technical communication. *Journal of Business and Technical Communication, 23*(2), 174–215. https://doi.org/10.1177/1050651908329562

Samuel, S. (2019, November 15). How biohackers are trying to upgrade their brains, their bodies–and human nature. *Vox*. https://www.vox.com/future-perfect/2019/6/25/18682583/biohacking-transhumanism-human-augmentation-genetic-engineering-crispr

Samsung Advanced Institute of Technology (n.d.). *Human augmentation*. https://www.sait.samsung.co.kr/saithome/research/researchArea.do?idx=1

Slane, A., Pedersen, I., & Hung, P. C. K. (2020). *Involving seniors in developing privacy best practices: Towards the development of social support technologies for seniors*. Office of the

Privacy Commissioner of Canada. https://www.priv.gc.ca/en/opc-actions-and-de-cisions/research/funding-for-privacy-research-and-knowledge-translation/completed-contributions-program-projects/2019-2020/p_2019-20_03/

Spence, R., & Jaworski, M. (n.d.). *Next eye prosthesis.* https://eyeborgproject.tv/

Spencer, G. (2019, October 9). *The art of augmentation: Human intelligence and artificial intelligence working together.* Microsoft: Stories Asia. https://news.microsoft.com/apac/features/the-art-of-augmentation-human-intelligence-and-artificial-intelli-gence-working-together/

Stelarc. *Zombies and cyborgs: the cadaver, the comatose, and the chimera.* (n.d.). http://stelarc.org/documents/zombiesandcyborgs.pdf

Sundar, S. S., & Lee, E-J. (2022). Rethinking communication in the era of artificial intelligence. *Human Communication Research, 48,* 379–385.

Tham, J. (2018). Interactivity in an age of immersive media: Seven dimensions for wearable technology, Internet of Things, and technical communication. *Technical Communication, 65*(1), 46–65.

Tham, J., Burnham, K., Hocutt, D. L., Ranade, N., Misak, J., Duin, A. H., Ped-ersen, P., & Campbell, J. L. (2021). Metaphors, mental models, and multiplicity: Understanding student perception of digital literacy. *Computers and Composition, 59.* https://doi.org/10.1016/j.compcom.2021.102628

Thayer, K. (2014). Mapping human enhancement rhetoric. In S. J. Thompson (Ed.), *Global issues and ethical considerations in human enhancement technologies* (pp. 30–53). IGI Global.

Wachowski, L., & Wachowski, L. (Writers & Directors). (1999–2003). *The Matrix* [Movie series]. Warner Bros.

Walton, R., Moore, K. R., & Jones, N. N. (2019). *Technical communication after the social justice turn: Building coalitions for action.* Taylor & Francis.

Warwick, K. (2008, April 14). *Kevin Warwick: Cyborg life* [Video]. Infonomia. Youtube. https://www.youtube.com/watch?v=RB_l7SY_ngI

Weiser, M. (1991, September). The computer for the 21st century. *Scientific American,* 94–104.

Welcome to the ALTAI portal! (2020). European AI Alliance. https://futurium.ec.europa.eu/en/european-ai-alliance/pages/welcome-altai-portal

Wolfe, Cary. (2010). *What is posthumanism?* University of Minnesota Press.

Wu, J. (2019, December 17). Empathy in artificial intelligence. *Forbes.* https://www.forbes.com/sites/cognitiveworld/2019/12/17/empathy-in-artificial-intelligence/

Zoomin. (2020, September 15). *The future of technical communication (ft. Rob Gillespie)* [Video]. YouTube. https://www.youtube.com/watch?v=mwLTdmjdNfI

Zraick, K. et al. (2019, June 13). That sleep tracker could make your insomnia worse. *New York Times.* https://www.nytimes.com/2019/06/13/health/sleep-tracker-in-somnia-orthosomnia.html

2

DIMENSIONS, SCOPE, AND CLASSIFICATION FOR AUGMENTATION TECHNOLOGIES

Overview

The scope of human augmentation technologies needs to be defined through the complex contexts within which they emerge. The core of this chapter is identification of a changing set of value systems that constitute augmentation technologies as social phenomena, namely, beliefs surrounding enhancement, automation, and efficiency. The chapter specifically articulates and examines cognitive, sensory, emotional, and physical enhancements as a range of sub-categories for identifying the underlying values contextualizing the future of augmentation technologies. Chapter 2 illustrates how the Fabric of Digital Life (https://fabricofdigitallife.com) archive's metadata fields serve as a means to examine augmentation technologies and the complex socio-technical tradeoffs that technical and professional communicators navigate as they work to develop usable content and direction.

Key Questions

- How might the scope of human augmentation technologies be defined through the complex social contexts within which they emerge? How might augmentation technologies be identified as social, rhetorically motivated phenomena as they enhance human capabilities or productivity by adding to the body (or ambient environment around the body) cognitive, physical, sensory, and/or emotional enhancements?
- Technical and professional communicators (TPCs) need to plan for and be part of the design, adaptation, and adoption processes surrounding augmentation technologies. How might TPCs build understanding of the socio-technical complexities surrounding augmentation technologies?

DOI: 10.4324/9781003288008-3

- TPCs can examine the Fabric of Digital Life's (https://fabricofdigitallife. com) rich metadata fields to search for terms and technologies as a means to better understand augmentation technologies and the potential tradeoffs related to embodiment and usability. How do these dimensions and exemplary technologies bring new challenges in terms of TPC understanding audience, data, privacy, intellectual property, usability, and content management?

Chapter 2 Links

Throughout the chapter, we refer to articles, videos, and reports which can be found in a related chapter collection at Fabric of Digital Life called *Dimensions, scope and classification for augmentation technologies* (Duin & Pedersen, 2023). You can find a link to this collection at https://fabricofdigitallife.com.

Introduction

Computer science is oftentimes criticized for presenting technology in objective neutral terms, to the detriment of citizens in the public sphere. Kate Crawford, scholar of social and political implications of artificial intelligence (AI), writes that AI "is not an objective, universal, or neutral computational technique…. Its systems are embedded in social, political, cultural, and economic worlds, shaped by humans, institutions, and imperatives that determine what they do and how they do it" (2021, p. 220). Augmentation technologies are deliberately designed to improve the way people think, feel, and act. They are inherently informed by the value-laden dialogue surrounding them, all of which frames how they are designed, adopted, and adapted by society. Consequently, the scope of human augmentation technologies needs to be defined through the complex social contexts within which they emerge (Pedersen & Duin, 2021).

Fabric of Digital Life

We decided to adopt a unique methodology for analyzing augmentation technologies. Central to this book is its integration with Fabric of Digital Life ("Fabric," https://fabricofdigitallife.com/), a structured content repository for conducting social and cultural analysis about emerging technologies. Growing in content since 2013, Fabric (see Figure 2.1) provides a public, collaborative site for analyzing technological change and the social implications that arise during human adaptation over time (Pedersen & Baarbé, 2013; Pedersen & DuPont, 2017; Iliadis & Pedersen, 2018; Pedersen et al., 2020; Duin et al., 2021). Fabric uses the CollectiveAccess content management system, which implements the Dublin Core™ metadata initiative. It follows modes of technological invention over time through its corpus of videos, texts, and images.

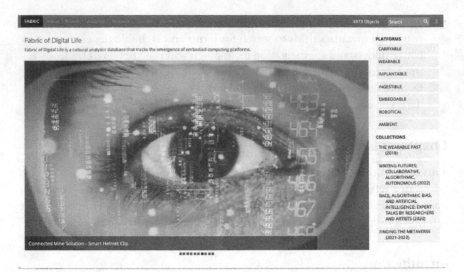

FIGURE 2.1 Fabric of digital life home page. Used with permission of Isabel Pedersen.

Fabric in its entirety is quite large, housing more than 5000 artifacts. However, it also houses thematic research collections on specific topics by curators who are often scholarly experts in a field.

Carole Palmer (2004) describes this task of "producing circumscribed collections, customized for intensive study and analysis in a specific research area." The novelty for collections is that "in many cases these digital resources serve as a place, much like a virtual laboratory, where specialized source material, tools, and expertise come together to aid in the process of scholarly work and the production of new knowledge" (Palmer, 2004). She explains the value of using digital collections for collaboration:

> The thematic collections concentrating on contextual mass and activity support are coming closest to creating a laboratory environment where the day-to-day work of scholars can be performed…. The most successful of these sites will move beyond the thematic focus to provide contextual mass and activity support that is not only responsive to what scholars currently do, but also to the questions they would like to ask and the activities they would like to be able to undertake.
>
> In the sciences the virtual laboratory, or collaboratory, concept has been around for some time…. Collaboratories are designed as media-rich networks that link people to information, facilities, and other people (Finholt, 2002). They are places where scientists can obtain resources, do work, interact, share data, results, and other information, and collaborate.

Fabric provides the tools to publish collections geared to specific research issues, and it also hosts interfaces for different archivists operating at different levels of research expertise and database proficiency to encourage the kind of collaboration Palmer describes.

Fabric Keywords and Metadata

Throughout each section in this book, readers can access more detail about each technology discussed by examining keywords and other metadata, discussed below. Each metadata category functions to categorize an aspect of a cultural artifact that is relevant.

Here we have chosen one artifact, a video about a type of augmentation technology, in order to explain how it has been archived in Fabric with a running commentary on the selections made. We ask that the reader consider it a "walk-through" of how this artifact was archived in Fabric using the metadata categories. It also provides some of the background on why it is culturally important as an artifact to be included in this book. This section demonstrates how Fabric can archive not only representative technologies, but also representative themes, opinions, events, and parties contributing to augmentation technology as an evolving phenomenon.

The example artifact "Beyond bionics: how the future of prosthetics is redefining humanity" (see Figure 2.2) is a 15-minute news broadcast for *The Guardian* (Sprenger, 2018). The description reads:

FIGURE 2.2 Fabric artifact and metadata for "Beyond bionics: how the future of prosthetics is redefining humanity." Used with permission of Isabel Pedersen.

> Bionic technology is removing physical barriers faced by disabled peo-
> ple while raising profound questions of what it is to be human. From
> DIY prosthetics realised through 3D printing technology to custom-
> ised AI-driven limbs, science is at the forefront of many life-enhancing
> innovations.
>
> *(Sprenger, 2018)*

The segment covers several types of augmentation technologies and interviews
people about the movement to enhance humans with technology, alongside
the stories of people who have adopted prosthetics for many different reasons.
It includes interviews with inventors as they explain their intent and motives
to create technologies such as brain-computer interfaces, digital prosthetics,
and other forms of cybernetics. The video clip also explores human evolution
through comments about "the future of our species," whether or not people
should be able to decide to upgrade human bodies, "redefine humanity," and
the ethical and moral debates surrounding these issues. Relevant for this book
is the manner through which this one clip produces overlapping persuasive
messages promoting augmentation technologies through the voices of different
people. It serves as an example of the normalizing of augmentation technologies that we are describing in this book.

Creators, Contributors and Publishers

Fabric follows *creators*, i.e., people, inventors, companies, news sources, governments, or entities primarily responsible for making the artifact. This data field
provides a way to follow inventors or writers over time. To get a better understanding of the new innovation, Fabric also follows *contributors*, i.e., people who
contribute to an artifact, including users. *Publishers* include entities that prepare
and make artifacts available to the public through a communication channel.

There are several interested parties informing this technology, and the archivist seeks to capture their contributions. In this example, we recognize the
journalist Richard Sprenger as the main creator of *Beyond Bionics* (Sprenger,
2018), as well as several content contributors, including Angel Giuffria, who
wears a prosthetic arm with enhanced physical capabilities; she discusses her
experiences and expertise on the cultural adaptation to these technologies. Her
messages might be aligned with the post-disability movement (Time, 2021),
which includes people like Victoria Modesta, who is an artist, musician, disability rights advocate, and creative practitioner who promotes bodily augmentations (Jones, 2017).

Dr. Miguel Nicolelis, an augmentation technology expert, is interviewed
in the *Beyond Bionics* news video for his brain-computer interaction inventions, with a display of his mind-controlled exoskeletons and wheelchairs inside his Duke University lab. He also discusses his belief that mind-controlled

technology will be significant now and in the future, noting that humans are "the first species capable to influence its own evolution" (Sprenger, 2018). This artifact contributes to a growing list of items by this inventor across several discursive types. Miguel Nicolelis is associated as a *creator* or *contributor* for several artifacts in Fabric, the first being a 2008 *Scientific American* article titled "Controlling Robots with the Mind." Therefore, brain-computer interaction is a cognitive enhancement technology that can be monitored over time across these artifacts. Nicolelis's comments in this artifact add to a portrait of his contributions to augmentation technologies that can be gleaned through multiple artifacts.

Media Type

Fabric classifies types of media. Media types include *newspaper articles, corporate advertising videos, film clips*, and many other categories of digital and print types.

Media genres inform how messages are delivered to audiences. *Beyond Bionics* is a video news broadcast, a journalistic piece tasked with publicizing how new technologies are leading to augmentation as a movement that is embraced by citizens and workers. The editorial expectations are different from a written news piece; *Beyond Bionics* proves points through realistic visual depictions. More like a short documentary, *Beyond Bionics* attempts a neutral gaze upon people who have adopted the technology, giving them a voice to express their choices and subjective opinions. It also features exciting visuals, another expectation of the medium.

Persuasive Intent

Closely related to media type, Fabric includes a field to specify the rhetorical or persuasive intent for an artifact that is explaining a technology. The subcategories are *academic, advertisement, art, entertainment*, and *information*.

New media, technology inventions, and innovations are usually conveyed through multiple categories representing the voices of different people, communities and organizations. Researchers present findings on inventions at *academic* venues to persuade an academic audience before inventions are sold to companies who then *advertise* those technologies years later for consumers. Adoption and adaptation to emerging technologies is tied to these kinds of persuasive acts.

Beyond Bionics is tagged as *information* because it is a segment of broadcast journalism from *The Guardian*, produced by the Guardian Media Group (GMG), a British-based media company, the publisher. The journalist has leeway to ask critical questions. For example, Sprenger asks Miguel Nicolelis if testing on monkeys is really necessary, pointing to the ethical issue of using lab animals for human augmentation technology development (Sprenger, 2018).

HCI Platform

Fabric collects technologies that are in physical or ambient contact with human subjects, *embodied computing platforms*. It concentrates on human-computer interaction platforms specifically designed for humans in movement, their lived reality: *carryable, wearable, implantable, ingestible, embeddable, robotical,* or *ambient* (see Figure 2.1).

By monitoring this range of technologies, Fabric of Digital Life is conducive to the study of human augmentation. *Beyond Bionics* is tagged as *embeddable* because it includes devices that add to or replace body parts. It discusses prosthetic devices through the voices of people who use them or inventors who make them. One theme is the manner through which "amputees are empowered and enhanced" (Sprenger, 2018) by embeddable devices that include myoelectric hands and other prosthetics. Another theme is that embeddable devices are a choice for human physical enhancement, pointing to the "beyond bionics" theme.

Discursive Type

There are three discursive types in Fabric: (1) *Inventions,* which are usually produced by an inventor or team, intending to explain, legitimize, or sell a technology; (2) *Responses to inventions,* which report on inventions by another source, including journalism, video lectures, or academic journal articles authored by non-inventors; and (3) *Objects of allusions,* which involve fictional technologies depicted in cultural artifacts, such as films, novels, or video games. Augmentation technologies are often described through the language of science fiction or speculative fiction.

People's comments and opinions function through discourses. "Discourse is a form of social action" (Van Dijk, 2001, p. 353); it indicates the roles of actors and intended social order through its instantiations in texts within context. *Beyond Bionics*'s discursive type is *Response to inventions*. It is produced by a large media enterprise (*The Guardian*), and audiences expect to learn more about an invention in a short but entertaining segment. The clip covers many different types of augmentation technologies to argue the theme that humans are being redefined by technology emergence. Interviews from two university researchers are intermeshed with interviews by three people and their families adopting and adapting to augmentation technologies, contributing to the legitimacy of the segment. The discourse predetermines our expectations over how ideas and people are to be represented in the text.

Related Body Part

This content field provides a researcher with the ability to access all artifacts relating to a particular part of a human body, for example *wrist* technologies, such as smartwatches. Most Fabric artifacts are tagged according to internal and external parts of bodies.

Beyond Bionics includes three people who use prosthetic arms, including Jamie Miller, a child whose father 3D prints them at home. The artifact is tagged with *arm* (related body part). This metadata category leads to this artifact being classified along with other technologies involving *arms*.

Keyword Categories

There are four ways to classify artifacts under subject area keyword categories: *Technology, Marketing, Augments*, and *General*.

Technology Keywords. This category includes technologies (hardware, software, and materials) that are directly mentioned or visually depicted in artifacts, such as *augmented reality (AR), AI*, or *brain-computer interface (BCI)*. Archivists also add Technology Keywords to help summarize a technical area by researching it, drawing on contexts relevant to artifacts.

Marketing Keywords. This category helps people track the commercialization of these technologies through the research and eventual creation of consumer products. It also follows companies, organizations, and other entities hosting or sometimes funding inventions. It includes keywords such as *Siri, Spot, Google Glass*, and *Sophia*.

Augments Keywords. This category classifies human capabilities that an artifact claims it could augment, enhance, diminish, or alter. We use the gerund grammatical form to designate these terms, for example *thinking, knowing, walking, learning, writing*, and *evolving*.

General Keywords. General keywords are collected in broader subject areas, such as *economics, health, disabilities, education*, or *military*. They give the visitor a general idea of the subject area and implications and connections to related fields.

Augmentation technologies intersect with multiple issues that inform their emergence. *Beyond Bionics* names and explores *transhumanism*, one of the topics central to this book and discussed in Chapter 1. Fabric has more than three hundred artifacts tagged with *transhumanism* as a general keyword. Sprenger asks one participant about what kind of arm "do you think you might have in ten years time?" The video also queries "the age of the cybernetically enhanced" and "the idea of voluntarily replacing limbs" (Sprenger, 2018). Angel Giuffria says:

> Ethics, when it comes to bionics, robots, AI, all these things are going to be huge… we need to make sure that we are prepared for these kinds of issues because we are not going to be ready for it, and then it is going to happen…. People that have the need to make these modifications to their bodies…. [we need to ensure that they will] have the choice.
>
> *(Sprenger, 2018)*

Giuffria touches on innovation, identity, governance, and ethics in these few statements, bringing relevant debates to the fore.

Publication or Creation Date

Each item has a publication or creation date. Most artifacts are sourced in the public sphere, such as videos posted or news sources published. The creation date is significant when artifacts are not published but sourced in other ways. If we record a historical or museum artifact, we name a creation date. Other important dates are *Date archived* and *Last edited*. These are useful because artifacts may be updated often by different archivists, the people responsible for adding content to the database.

These categories can be used to mark and understand the ambiguity surrounding augmentation technology. In the next section we provide in-depth discussion of augmentation technology dimensions.

Augmentation Technology Dimensions

Throughout this book, we discuss myriad applications through the lens of people and rightsholders using the terms from different points of view. Our intention is to provide a means to classify augmentation technologies in socio-technical terms.

Different sectors use the term "augmentation technology" to explain, justify, and promote the adoption of new technologies; however, they do not use the term consistently. We conducted a search of the Association of Computing Machinery (ACM) digital library using the terms "augmentation technologies," "human augmentation," and "enhanced humans" and found 264 articles with varying degrees of relevance. This initial search prompted a more comprehensive full-text search on several bibliographic databases, ACM digital library, Engineering Village, IEEE, ScienceDirect, Scopus, and Web of Science. The keywords searched were "human augmentation" and "augmentation technology." This led to a corpus of 650 augmentation technology articles published between 2000 and November 2020 from the field of science and technology, informing this book as well as several supporting projects.

Augmentation technologies are embodying; they add to the body (or ambient environment around the body). Isabel Pedersen and Andrew Iliadis (2020) define embodied computing as those technologies that "exist in topographical [on the body], visceral [in the body], and ambient [around the body] relationships with the body" (p. xi). For the sake of definition, we identify a set of value systems that frame augmentation technologies as social phenomena, namely, enhancement, automation, and efficiency:

- Enhancement—Enhancement technologies have been defined by the Sienna Project as "a modification aimed at improving human performance and brought about by science-based and/or technology-based interventions in or on the human body" (Jensen et al., 2018, p. 9). We identify physical, cognitive, sensory, and emotional enhancement as a range of subcategories (see Table 2.1). These subcategories help us classify nearly all types of augmentation. More importantly, these provide a working terminology for TPCs that allows for discussion of the intent to augment a process without being limited to discussion of a specific piece of hardware. Socio-technical concepts are as important as technologies.
- Automation—We assert that automation has been so celebrated and promoted extensively in the past decade that it has changed the playing field for our expectations. Augmentation technology assumes that a degree of automation is always valuable, a logic imposed on it largely by the mass adoption of AI. We intend to unpack this assumption as a somewhat invisible rhetoric used to drive the development and adoption of these technologies.
- Efficiency—Augmentation technologies are geared to the value of efficiency. Humans, human work, and human experiences are reconceptualized in terms of usefulness or avoiding wasting energy, time, or resources. Here we note a shift in this paradigm. While physical enhancement might have always assumed efficiency as a goal, enhancement in thinking, sensing, and feeling is new territory in discursive deployments, and it may lead to harms that have gone uninvestigated.

The burgeoning augmentation technologies market masks complex socio-technical tradeoffs that emerge from value framing, the intent to enhance, automate, and build efficiency. It can also mask inequities, biases, and power imbalances experienced by intended users.

TPCs are encouraged to understand these value systems as part of their work to develop usable and ethical content. Table 2.1, Scope of Augmentation Technologies, provides an overview of the four enhancement subcategories that we use to classify augmentation technology. We provide examples of types of augmentation technologies, including emerging hardware, software, and applications. For each category, we provide a set of synonyms used across different sectors when defining the augmentation as well as the general goals for its use. For goals, justifications, or rhetorical rationale, we summarize the intent that motivates the design of a technology.

Cognitive Enhancement

Goals for cognitive enhancement include the perceived need to be smarter, be more knowledgeable, think faster, remember more, know more, learn faster, learn more efficiently, edit dysfunctional memory, or be more reasonable.

TABLE 2.1 Scope of augmentation technologies

Scope	Cognitive enhancement	Sensory enhancement	Emotional enhancement	Physical enhancement
Types of augmentation technologies including emerging hardware, software, and applications	Brain–computer interaction Human–AI interaction/ collaboration Explainable AI AI Virtual assistants	AR Virtual reality Extended reality (XR) 3D immersive space Smart contact lens Cochlear implants (Auditory) smart translation Haptics Garment-integrated tactile displays	Affective computing Emotion detection Emotion decoding technology Biometrics Artificial Emotional Intelligence (Emotion AI)	Exoskeletons Exosuits Smart shirts Programmable fabric Digital fabric E-textiles Wearable robotics Smart prosthetics
Synonyms	Intelligence amplification Memory amplification Machine augmented intelligence Enhanced intelligence Neurostimulation technology Neural prostheses Assisted reasoning and decision-making Enhanced rationality	Perceptual enhancement Immersive enhancement	Affective enhancement Emotion manipulation Emotion regulation	Superhuman technology Assistive technology
Goals, justifications, or rhetorical rationale	be smarter think faster be more reasonable be more knowledgeable know more learn faster learn better (efficiently) remember more edit memory	experience more through manipulation of the senses augment one's senses to construct an enhanced reality escape one's sensory reality focus better	feel more and different emotions feel less emotion control emotion be happier	exceed human physical abilities be stronger be faster be safer live longer live healthier work harder work longer endure more work more efficiently play sports better
Platforms (Fabric integration)	Wearable Implantable Robotical	Wearable Implantable Ambient Robotical	Wearable Implantable Robotical	Wearable Ingestible Embeddable

These terms are always changing, but they cover concepts related to human cognitive abilities. Regarding *thinking* enhancement, in the 1960s Doug Engelbart explored how cognitive abilities could be enhanced through rapid access to information and increasing user comprehension, developing an early user interface for intelligence amplification (1962). As Danry and colleagues (2020) at the MIT Media Lab note, researchers have developed and studied various interfaces for cognitive enhancement (Maes, 1995; Starner, 2002; Huber et al., 2018) and group decision-making (Valeriani et al., 2019). There are countless examples of cognitive enhancement and the synonyms used to promote it. Currently, the explosion of public debate surrounding large language models and applications like OpenAI's ChatGPT informs the discourse of cognitive enhancement. The speculation implies that anyone who wants to use the tool can master writing, a capability that was always considered to be a craft of human expertise.

Throughout this book, we address ways that cognitive enhancement is described, proposed, or even justified to help our readers understand how it progresses and to recognize its impact on society and professions under states of emergence. Whether a proof-of-concept or product on the market, these kinds of technologies can be viewed in terms of how they promise or claim to augment a human. In this category, we are looking for specific claims that a technology will help someone be smarter, more knowledgeable, think faster, remember more, know more, learn faster, learn more efficiently, edit dysfunctional memory, or be more reasonable. These terms are always changing, but they cover concepts related to human cognitive abilities.

According to Gartner Inc.'s glossary, "Augmented intelligence is a design pattern for a human-centered partnership model of people and artificial intelligence (AI) working together to enhance cognitive performance, including learning, decision making and new experiences" (Gartner, n.d.). In this way, cognitive enhancement comes about as a result of partnership in which people and AI work together. Instead of a "design pattern," Wigmore (2019) defines augmented intelligence as "an alternative conceptualization of artificial intelligence," also focusing on AI's assistive role, and "emphasizing the fact that cognitive technology is designed to enhance human intelligence rather than replace it" (Wigmore, 2019). IEEE's Digital Reality publication also emphasizes that "augmented intelligence is designed to enhance, not replace," using "machine learning and deep learning to provide humans with actionable data" (IEEE Digital Reality, n.d.).

Judith Hurwitz and colleagues (2020) focus on the business power of this human-machine collaboration, emphasizing that "humans need to evaluate the results of the automated tasks, make decisions in non-routine situations, and also assess if and when the data must be changed due to changing business needs and demands" (p. 2). They discuss "weak versus strong augmentation": automating tasks done by humans is weak augmentation and results in making the tasks more efficient and cost effective; strong augmentation combines machine with human assessments. They emphasize that "the goal of augmented intelligence is to enable human–machine collaboration to produce positive outcomes that neither machines nor humans alone can achieve," with the steps needed to implement augmented intelligence being as follows:

- Decide whether to change the business process and task flow for human-machine cooperation to drive better outcomes;
- Select which tasks and decisions within the business process to automate;
- Determine the proper AI tools;
- Determine what data to acquire to better understand and model the business and customers;
- Build the data models; and
- Test the results for reliability and accuracy (p. 3).

More recently, Valdemar Danry and colleagues (2020) built on this research to develop Wearable Reasoner, an AI assistant device for enhancing judgment and decision-making in real-time, encouraging people to be more skeptical and to seek and demand evidence before jumping to conclusions (see Figure 2.3). The device uses smart glasses (Bose Frames) with open-ear speakers that enable private audio feedback while not blocking the ear canal. This device has a microphone that detects utterance input and it can be connected via Bluetooth to a smartphone for additional processing through a mobile application. This wearable, audio-based system examines logical structures of an argument and offers real-time evaluative and analytic feedback to the wearer through its use of "an algorithm capable of delivering relevant feedback on argumentative structures" (Danry et al., 2020, p. 4).

In the experimental design study, participants used one of three versions of the device: one equipped with explainable AI that responded with detail

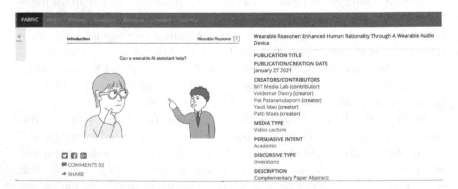

FIGURE 2.3 Fabric artifact for the Wearable Reasoner. Used with permission of Isabel Pedersen.

on the claims and evidence heard; one equipped with only non-explainable AI, so it responded with simply "supported" or "non supported" to a statement heard; and a control device with no AI response (Danry et al., 2020). Their results show the explainable AI helped users to "differentiate between statements supported by evidence and without. When assisted by an AI system with explainable feedback, users significantly consider claims supported by evidence more reasonable and agree more with them compared to those without" (Danry et al., 2020, p. 1). Obvious concerns as this technology develops and becomes mainstream include overreliance on the device and influence of the algorithmic system on judgments. It justifies its design strategy on human shortcomings and susceptibility "to manipulation," using automation for corrective purposes for humans who need to be more rational (AHs—Augmented Humans International Conference, 2021). At the same time, it also promotes a symbiotic relationship with AI "cognitive symbiosis" suggesting that cognitive enhancement should consist of dialogic exchange, rather than a purely authoritative relationship.

What human capacities is Wearable Reasoner aiming to augment? To monitor how this invention is associated with cognitive enhancement in terms of social or even ontological phenomena, we added this invention to Fabric of Digital Life and displayed some of the item metadata (see Figure 2.4). We offer this breakdown of Wearable Reasoner (AHs—Augmented Humans International

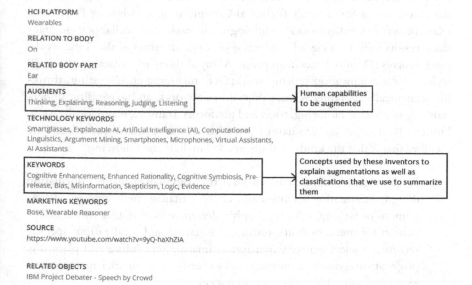

FIGURE 2.4   Metadata for "Wearable Reasoner: Enhanced Human Rationality Through A Wearable Audio Device." Used with permission of Isabel Pedersen.

Conference, 2021) as an exemplar for classifying cognitive enhancement. In this example, we assign human capabilities in the **Augments** field using the concepts raised in the inventors' paper and video about this invention. In doing this, we create a portrait of claims about augmentation. Wearable Reasoner promises the capacity to think better, and improve reasoning in order to make better judgments, hence we assign *thinking, explaining, reasoning, judging,* and *listening.* In the **Keyword** category, we classify this team's usage of *enhanced rationality* to characterize cognitive enhancement. The **Technology Keywords** field is also important because it places *AI, explainable AI, AI Assistant,* or *argument mining* as having the agency, sophistication, functionality, and speed to achieve the augmentation for humans (**Augments** terms). As we classify augmentation technologies in these human-centric terms, rather than relying on only technical terms (machine-centric), we ground our analysis with a terminology better suited to social domains.

Additional companies deploying augmentation technologies include Augmedix, Inc., which has partnered with Google to use Google Glass coupled with AI capabilities to perform automated remote note-taking and compare patient symptoms with millions of others to predict effective treatment (AI Med., 2021); and Boeing, which has deployed AR glasses to guide technicians as they wire planes, reducing error rates by 25% (Upskill, n.d.). While each of these may increase efficiency and productivity, emerging technologies such as Wearable Reasoner promise a deeper level of cognitive enhancement, a powerful realm for TPCs. In a recent study of 20 TPC industry leaders—members of the University of Minnesota's Technical Communication Advisory Board—to examine writer identity, socio-technological literacies, and collaboration practices, results indicate a clear broadening of TPC identities as the TPC workplace evolves (Duin & Breuch, in press). Many abilities are critical to the 2020 technical communication writing workplace, including collaborating, thinking strategically, building relationships and networks, and expanding understanding of content authoring, tools and platforms, translation and localization, business ROI, legal and regulatory compliance, and usability/audience. One member shares that the kinds of skills needed include the following:

> Requirements analysis, customer relationship management, information design, information architecture, content management, content development or writing, editing, graphic design, system testing of the information for users, usability testing, translation and localization, specific technical subject matter expertise, estimating scheduling and planning, project management, authoring tool expertise, content management tool expertise, and information maintenance.

Another member notes a critical literacy around content strategy as being the ability to define a means by which to say yes or no to the next plausible idea that comes along, and others mention increased reliance on algorithmic tools to keep track of the exploding amount of information. Wearable Reasoner is being positioned to assist.

Sensory Enhancement

Goals for sensory enhancement include the desire to experience more through augmentation of the senses, to augment or reconstruct one's reality through digital representations, to escape one's sensory reality, and to focus better. Sensory enhancement has a long history of both personal and industrial non-digital uses: glasses, binoculars, microscopes, stethoscopes, hearing aids, and welder's goggles augment senses functioning as an aspect of one's self. Hololens 2, a mixed reality smart glass device, aids users in viewing intricate details of 3D images, and more recently, eSight is promoted as a "life-changing device for people with low vision." This wearable device contains cameras on the front that provide information about the near-eye quality environment and depict it on a screen that sits right in front of the wearer's eyes. Articles about eSight point to the kinds of discourses we have been discussing. Featured at the 2018 Augmented World Expo™, eSight is described as "AR for the blind… straight out of Star Trek" by science accommodators using popular science fiction to frame adoption for this sensory augmentation (Nichols, 2018).

We use the term "immersive technologies" to mean forms of sensory augmentation that are often visual but can also include auditory or tactile immersion. They include virtual reality, AR, mixed reality, and immersive videos. Traditionally, VR immerses users in a completely computer-generated 3D environment, while AR combines virtual components as an overlay over the real-world, a sensory augmentation. Justifications for immersive augmentation are myriad. The idea of a Metaverse is important to this sensory enhancement category. It signifies actual immersive technologies converging as if they are one phenomenon, hype about it by Big Tech commercial backers, and the anticipatory assumption that Metaverse is a moment in time: *a future that will happen*. Much immersive business technology undergoes the early stages of emergence due to the expectation that Metaverse will happen, a coordinated effort to dramatically evolve the internet.

Sensory enhancement combined with AI promises to advance human abilities significantly, pointing to the posthuman discourse that now informs expectations for efficiency and automation of human behaviors. Inventor Arnav Kapur says that using his AlterEgo device will help people to become "better at being human," cloaking it with the augmentation rhetoric that defines the field (2019). AlterEgo is a wearable AI device "with the potential to let you silently talk to and get information from a computer system, like a voice inside your head" (2019). It enhances *speaking* by transcribing "words that the user verbalizes internally but does not actually speak aloud" (Hardesty, 2018). It is a "silent speech" augmentation that listens to the user by sensing the neuromuscular signals of the jaw, and responds by conduction through the user's bone so that it can only be interpreted by the wearer (Hardesty, 2018). AlterEgo also enhances *listening*. One can silently query the AlterEgo AI agent for information from a source and receive audio answers on the headset. We classify this invention as a manipulation of the senses, but one can imagine the evocative ways that it might augment *thinking* as well.

What human capacities is AlterEgo aiming to augment? To monitor how this invention is associated with sensory enhancement in terms of social phenomena, we added this invention to Fabric of Digital Life (see Figure 2.5) and displayed some of the item metadata (see Figure 2.6).

We offer this breakdown using AlterEgo (Kapur, 2018) as an exemplar for classifying sensory enhancement. We assign human capabilities in the **Augments** field using the concepts raised in the inventors' video through verbal explanation or visual depiction. AlterEgo denotes an augmentation to *listening* and *speaking*, but it also asks us to consider the connotations implied with speaking silently. It adds a further layer: *what if one could silently have a dialogue with an AI agent/expert?* Given the opportunity to experience one's senses in this way, AlterEgo is tagged with the **Technology Keyword** *augmented reality*. Closely twinned with cognitive enhancement, AlterEgo's benefits include leveraging one's inner thoughts with a (silent) voice and means to communicate. One of the inventors' goals is stated overtly: "AlterEgo aims to combine humans and computers—such that computing, the internet, and AI would weave into human personality as a 'second self' and augment human cognition and abilities" (Project AlterEgo, n.d.). We use several **Keywords** to classify this highly

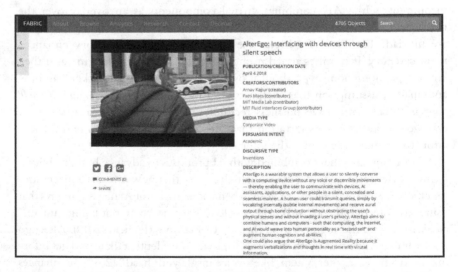

FIGURE 2.5   Fabric artifact for "AlterEgo: Interfacing with devices through silent speech." Used with permission of Isabel Pedersen.

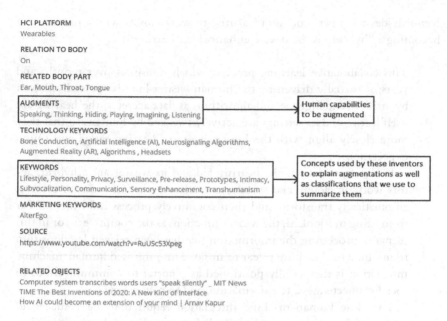

HCI PLATFORM
Wearables

RELATION TO BODY
On

RELATED BODY PART
Ear, Mouth, Throat, Tongue

AUGMENTS
Speaking, Thinking, Hiding, Playing, Imagining, Listening → Human capabilities to be augmented

TECHNOLOGY KEYWORDS
Bone Conduction, Artificial Intelligence (AI), Neurosignaling Algorithms, Augmented Reality (AR), Algorithms , Headsets

KEYWORDS
Lifestyle, Personality, Privacy, Surveillance, Pre-release, Prototypes, Intimacy, Subvocalization, Communication, Sensory Enhancement, Transhumanism → Concepts used by these inventors to explain augmentations as well as classifications that we use to summarize them

MARKETING KEYWORDS
AlterEgo

SOURCE
https://www.youtube.com/watch?v=RuUSc53Xpeg

RELATED OBJECTS
Computer system transcribes words users "speak silently" _ MIT News
TIME The Best Inventions of 2020: A New Kind of Interface
How AI could become an extension of your mind | Arnav Kapur

FIGURE 2.6 Metadata for "AlterEgo: Interfacing with devices through silent speech." Used with permission of Isabel Pedersen.

ambitious intent, including that *lifestyle, personality,* and *privacy* would all be enhanced or altered. The idea of "weaving" the internet and AI into one's personality to better oneself—combining humans and computers—is a transhumanist motive, hence the keyword *transhumanism.* As a research prototype, AlterEgo is depicted mostly in personal scenarios, shopping, internet searching, or playing chess, rather than industry settings. Interestingly, these inventors suggest that AlterEgo will better inform people's privacy, that silent speech and communication with human and nonhuman actors might be concealed from the public (Kapur, 2018). Yet data privacy or control over one's data is not discussed.

In *Writing Futures,* we highlight Krista Kennedy's (2017) study of everyday collaboration in her use of digital hearing aids in which she examined the design of interfaces and systems to enhance her hearing. She writes:

> By assuming that a collaborative and integrated relationship will develop between human and nonhuman, we can better design essential wearables as portals for social connectedness, information, convenience, and in the case of hearing aids, communication. Doing so requires moving beyond design processes that construct wearers and interfaces as separate agents with scheduled points of contact during early development and then not again until the usability testing stage.
>
> *(p. 41)*

Kennedy describes her process of learning to work closely with a machine, of becoming a "hybrid" as the device enhances her hearing:

> This collaborative learning process, which transpires over months and years, is partially driven by the human wearer but also partially driven by artificial intelligence and algorithms as data accretes, the hearing aid itself learns, and its settings are actively tweaked by humans so that they more closely align with the human wearer's hearing range and sound expectations....
>
> In this formulation, the hearing aid and its wearer are a hybrid, actors working closely and transformatively together in order to receive, automatically transform, and then cognitively process information. Far from being overlooked, the wearer functions as the cognitive actor in this scenario, processing the information that has been relayed by algorithm to machine to algorithm to ear to mind. This promised human/machine integration is rhetorically positioned as a portal to communication, to social connectedness, to information, and to convenience....
>
> The close human-machine integration required to be a successful wearer of these aids can lead to a collaborative relationship between human and machine as the wearer learns to work very closely with multiple machine agents that include hardware, directional mics, multiple processing algorithms, satellite telemetry, and an iPhone with an app that facilitates control of multiple functions as well as acts as a central element for geolocation telemetry.
>
> *(pp. 41, 44, 49)*

In sum, Kennedy experiences a transformative sensory enhancing scenario through a collaborative, human/machine integration.

Multi-sensory technology now also augments flavor perception to enhance human–food interaction design (Velasco et al., 2018), sensory marketing (Petit et al., 2015), and the arts (Vi et al., 2017). Valentina Cesari and colleagues (2021) propose use of multi-sensory enhancements as a means for learners to better engage behaviorally, emotionally, and cognitively with online learning. In addition, the US Army provides soldiers with "Tactical Augmented Reality" (TAR), improved situational awareness with the use of geo-registration AR technology that replaces both night-vision goggles and handheld GPS systems to determine positions, using an enhanced image of the target that is seen through the eyepiece (Vergun, 2017). AR glasses, AR use during sports and events, AR for improving learning outcomes, AR in medicine, AR in the car and for travel, and AR to enhance work applications are all expanding. At the CES 2021 show, AR and AI "stole the show" as Panasonic's AI-powered AR device helps users detect dangers while driving; Lenovo's ThinkReality A3 AR glasses pair up with smartphones and computers for immersive training;

and the Binah.ai system allows users to look at a phone camera for 15 seconds and then receive detail on 15 health indicators "at medical-grade levels of reliability" (Nextech, 2021).

Emotional Enhancement

Will AI ever know human emotions enough to be able to augment people's lived experiences with some form of emotional clarity? All over the world, researchers, companies, and even governments are working to create ways to interact with humans through "synthetic emotions" in AI applications. A great deal of excitement circulates in public popular science about AI agents that claim to read emotions. Most unique at the CES 2022 show appears to be the Best Innovation Award in Robotics, the MOFLIN AI pet (see Figure 2.7) capable of expressing emotions that change based on an algorithm that analyzes the surrounding environment through sensors. This device can tell when different people interact with it, reacting accordingly. It is even referred to as having "emotional capabilities" (Vanguard Industries Inc., 2022).

Goals for emotional enhancement of humans include the desire to feel that devices understand one's emotions, to control one's emotion, or even to be happier or more fulfilled. Countless domains now work toward ambitious goals to solve world problems upon the expectation that AI will someday achieve human-like empathy. Care systems for aging populations are to be automated if emotional augmentation can progress enough, for example. Critical to

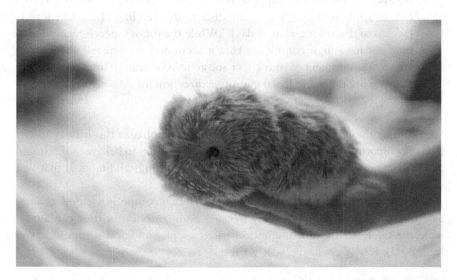

FIGURE 2.7 MOFLIN, an AI pet with emotional capabilities. Used with permission of Masahiko Yamanaka, Vanguard Industries, Inc.

emotional enhancement is that ethical design practices must underpin all work in this area to ensure that human rights are always respected.

A foundational person to the study of emotional enhancement technology is Rosalind Wright Picard, who is credited with starting the branch of computer science known as affective computing. Founder and director of the Affective Computing Research Group at the MIT Media Lab, Picard pioneered focus on social robots and wearable computers, working to "give computers the ability to recognize, understand, [and] even to have and express emotions" (Picard, n.d.). Earlier we mentioned the robot Rachel by Emoshape that uses algorithms to respond to meaning and language around her in real-time, and then express herself accordingly (Emoshape Inc., 2018). At the time of her groundbreaking book *Affective Computing* (1997), Picard asked readers to envision a robot entering your kitchen as you prepare for guests. Here we alter Picard's description slightly [see inserts throughout] to envision a robot entering one's technical and professional communication workspace:

> Imagine your robot entering [your TPC workspace as you prepare for a virtual usability test]. The robot looks happy to see you and greets you with a cheery "Good morning." You mumble something it does not understand. It notices your face, vocal tone, [cluttered screen], and your slamming [your hands on the desk], and infers that you do not appear to be having a good morning. Immediately, it adjusts its internal state to "subdued," which has the effect of lowering its vocal pitch and amplitude settings, eliminating cheery behavioral displays, and suppressing unnecessary conversation. Suppose you exclaim, "[Darn]!!" [peering at the lack of a robust connection to the distant sites], adding "I can't believe I [agreed to this testing work today]." While the robot's speech recognition may not have high confidence that it accurately recognized all of your words, its assessment of your affect and actions indicates a high probability that you are upset and [need cognitive direction for possible next steps].
>
> *(Picard, 2007)*

We illustrate these early design goals by Picard to emphasize this history in affective computing, that the desire to imbue our machines to behave as if human and to help, augment, or even care for humans is a longstanding goal that we will return to in later chapters.

At the same time, we acknowledge that affective computing is undergoing extensive criticism and controversy because companies claim they can automate the inference of human emotion based on physiological stimuli (Crawford, 2021). Kate Crawford (2021) writes:

> For the world's militaries, corporations, intelligence agencies, and police forces, the idea of automated affect recognition is as compelling as it is

lucrative. It holds the promise of reliably filtering friend from foe, distinguishing lies from truths, and using the instruments of science to see into interior worlds.

(p. 162)

However, systems detecting and analyzing facial expressions—automated affect detection systems—are being deployed in industry despite "a lack of substantial scientific evidence that they work" (p. 163). Likewise, the IEEE Global Initiative on Ethics of Autonomous and Intelligent Systems (2019) states the following:

> While A/IS [autonomous and intelligent systems] have tremendous potential to effect positive change, there is also potential that artifacts used in society could cause harm either by amplifying, altering, or even dampening human emotional experience. Even rudimentary versions of synthetic emotions, such as those already in use within nudging systems, have already altered the perception of A/IS by the general public and public policy makers.
>
> *(p. 90)*

We raise and address this controversy as part of the book's agenda to incorporate digital literacy, AI literacy, and other techniques to deal with the harm that becomes apparent as certain augmentation technologies emerge.

With focus on programming physical materials to promote well-being similar to human touch, Athina Papadopoulou and colleagues (2019), also at MIT, developed a programmable Affective Sleeve, a wearable device that produces warmth and slight arm pressure to promote calmness and reduce anxiety. They note that "the awareness of the physiological state of our own bodies is highly interrelated to our emotional awareness—our ability to process and express our emotions" (p. 305). They asked 18 MIT college students to wear the sleeve while taking a quiz to induce high stress levels. Students randomly received either an inactive sleeve (control group), a fast rhythmic haptic action sleeve (fast group), or a slow rhythmic haptic action sleeve (slow group). All students also wore a wristband sensor and a chest patch to record respiration and its variability. Results showed the pace of haptic action to influence both breathing rate and perception of calmness, with the slower device providing the greatest benefit.

What human capacities is Affective Sleeve aiming to augment? To monitor how this invention is associated with emotional enhancement in terms of social phenomena, we added this invention to Fabric of Digital Life (see Figure 2.8) and displayed some of the item metadata (see Figure 2.9).

We offer this breakdown using Affective Sleeve (Papadopoulou et al., 2019) as an exemplar for classifying emotional enhancement. We assign human

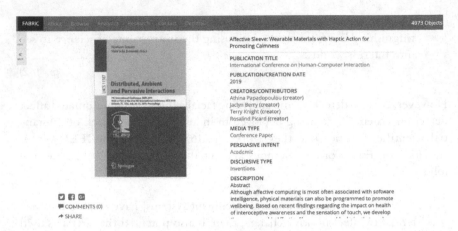

FIGURE 2.8 Fabric artifact for Affective Sleeve, published in N. Streitz and S. Konomi (Eds.), *Distributed, Ambient and Pervasive Interactions*. Used with permission of Isabel Pedersen.

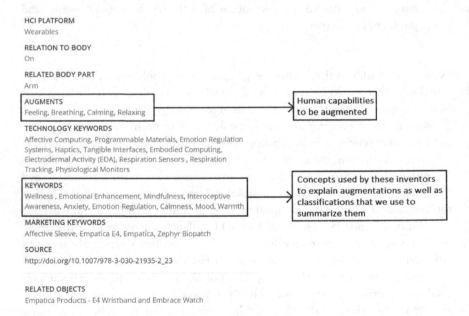

FIGURE 2.9 Metadata for the Affective Sleeve publication. Used with permission of Isabel Pedersen.

capabilities in the **Augments** field using the concepts raised in the academic article. Note that Affective Sleeve claims an augmentation to *feeling* and *calming*. These inventors use the **Technology Keyword** *programmable materials* and explain that while emotional enhancement is usually achieved through software, their use of *haptics* to control emotions is novel (p. 904). The **General**

Keywords *wellness* and *mindfulness* point to the discourses of self-care or the idea that people can control and regulate their own emotions, (i.e. *emotional regulation*).

Physical Enhancement

Goals for physical enhancement include the opportunity to be stronger, be faster, live longer, live healthier, work longer, physically work more efficiently, and play sports better. Fabric holds 160 artifacts on exoskeletons, sometimes called "wearable robotics," slated to dramatically enhance human capabilities proposed to combine with AI, Internet of Things, AR, and other technologies.

Pedersen and Mirrlees (2017) define an exoskeleton as "a body-worn human enhancement technology that takes the form of a suit, made to heighten, surpass, or add physical and cognitive abilities" (p. 39). They discuss how transhumanism, most often discussed in relation to posthumanism, "signifies a semantic middle ground between the categorical species named human and the scientifically or technologically evolved category known as posthuman" (p. 39). They identify four themes associated with transhumanist rhetoric (see Table 2.2): legitimate science, human agency, superheroism, and vulnerability; and they employ Burke (1950) to explain how each theme works to change people's attitudes toward physical enhancement technologies. They analyzed a corpus of texts from advertising, popular science, and science fiction films to understand the intertextual themes across this discourse as a way to identify socio-ethical beliefs about the technology as conveyed through this discourse. In doing so, they "make salient an irony" (p. 44): in our post-millennial, networked society, physical enhancements "represent a sign of societal progress and a sensationalized transhuman rhetoric of human improvement or utopian betterment in accommodated science" while "fictional exoskeletons frame a dystopian transhuman and oftentimes-militarized identity that instantiates vulnerability," revealing "defenselessness and fear rather than idealized liberal-humanist hope that we so commonly expect" (p. 44). While alluring, this friction "obscures our ability to identify ideology or how this technology is used to frame an artificial trans-human future" (p. 45).

Perhaps more easily understood are exoskeletons used in construction industries that provide metal frameworks with motorized muscles to multiply the wearer's strength (Thilmany, 2019). It is unclear how technical communicators will use exoskeletons for future physical enhancement, but these technologies are quickly being developed in response to human needs, and technical communicators are positioned with strategic knowledge as part of their design, adoption, and adaptation. For example, Naked Prosthetics, https://www.npdevices.com/, creates custom hand prosthetics to give users greater dexterity using finger prostheses. An emerging example includes 3D bioprinting in which materials are created that feel like bones, organs, and skin. 3D bioprinting is

TABLE 2.2 Themes associated with transhumanist rhetoric

Theme 1: Legitimate Science	Theme 2: Human Agency	Theme 3: Superheroism	Theme 4: Vulnerability
Legitimization of technological progress through science, technology, engineering, and math fields and practices. Collaboration between military industrial complex and university research hubs.	Rejection of human biological limitations to justify technological progress—e.g., humans should be stronger, fly, overcome disease, live forever, etc. Collaboration between medical and military industrial complex research.	Celebration of superheroism or allusion to fictional superheroes, and identity tactics used to frame the subject in relation to heroes. Collaboration between Hollywood and military industrial complex goals. Cinematic franchises accommodate military goals.	Reaction to a shared sense of vulnerability, militarism, and fear of the future, abstracted future outcomes rather than specific ones. Identity tactics used to make fear personal. Post 9/11 contextualization and militarism.

Source: © 2017 IEEE. Reprinted, with permission, from Pedersen and Mirrlees (2017).

envisioned to develop 3D artificial corneas to replace human donors, to print fully vascularized hearts, and to print skin cells directly on a burn wound. Similarly, artificial blood substitutes are being designed to better deliver oxygen in an emergency and to help across a large variety of syndromes, and nanobots are being deployed into the human body to target and attack disease. Overlapping with cognitive enhancement is an artificial chip known as Optogenetics, an augmented technology designed to mimic the human brain, making it super-efficient for storing, deleting, and processing crucial information (Rodriguez, 2019).

Yet another example is digital fiber designed at MIT that senses, stores, analyzes, and infers activity after being sewn into a shirt (see Figure 2.10). Professor Yoel Fink and colleagues, as shared by Ham (2021), describe this work as "the first realization of a fabric with the ability to store and process data digitally, adding a new information content dimension to textile and allowing fabrics to be programmed literally." Representing on-body AI, the fiber is able:

> to determine with 96% accuracy what activity the person wearing it was engaged in.... With this analytic power, the fibers someday could sense

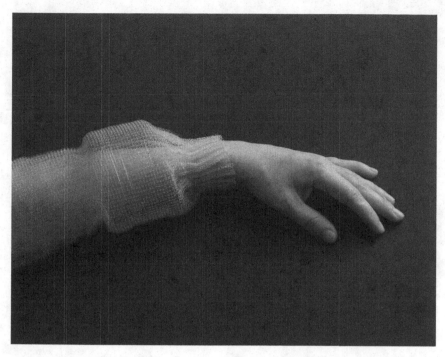

FIGURE 2.10 Digital fiber in a shirt. Textile design and development by Anna Gitelson-Kahn, photography by Roni Cnaani. Used with permission of Anna Gitelson-Kahn.

and alert people in real-time to health changes like a respiratory decline or an irregular heartbeat, or deliver muscle activation or heart rate data to athletes during training.

We offer this breakdown using this digital fiber invention as an exemplar for classifying physical enhancement (see Figure 2.11). We assign human capabilities in the **Augments** field by using the language described in one article about the invention (Ham, 2021), as well as our own observations. The technology will perform autonomous body and health monitoring, leading us to include caring, feeling, and monitoring as human capabilities that are augmented. One of the more interesting claims is that it could be used for inferring physical actions or predicting a human movement before it happens. The **Technology Keywords** include *AI* and *on-body intelligence* (a novel term), which are afforded the agency to back this innovation. This journalistic article announces a publication in *Nature* by the team at MIT, which goes into extensive detail about the invention. However, it offers little information about real-world application or why someone would want to self-augment with such a smart shirt. Much physical augmentation technology research is funded by militaries. This digital fiber

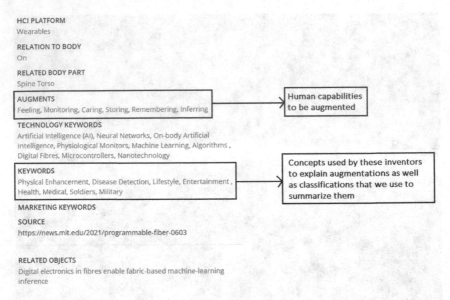

HCI PLATFORM
Wearables

RELATION TO BODY
On

RELATED BODY PART
Spine Torso

AUGMENTS
Feeling, Monitoring, Caring, Storing, Remembering, Inferring

Human capabilities
to be augmented

TECHNOLOGY KEYWORDS
Artificial Intelligence (AI), Neural Networks, On-body Artificial
Intelligence, Physiological Monitors, Machine Learning, Algorithms,
Digital Fibres, Microcontrollers, Nanotechnology

KEYWORDS
Physical Enhancement, Disease Detection, Lifestyle, Entertainment,
Health, Medical, Soldiers, Military

Concepts used by these inventors
to explain augmentations as well
as classifications that we use to
summarize them

MARKETING KEYWORDS

SOURCE
https://news.mit.edu/2021/programmable-fiber-0603

RELATED OBJECTS
Digital electronics in fibres enable fabric-based machine-learning
inference

FIGURE 2.11 Metadata for "Engineers create a programmable fiber" article from MIT News. Used with permission of Isabel Pedersen.

invention was supported by the US Army Institute of Soldier Nanotechnology, National Science Foundation, and the US Army Research Office. We add *militaries* and *soldiers* to the **Keywords** to signify the involvement of these entities.

Enabling Technology

International multi-stakeholder organizations and actors (e.g., WHO, WIPO, UN) orient augmentation technologies toward a collective global worldview for inclusive ends. We conclude this core section with discussion of the evolution toward mass use of enabling technology as a key finding in the World Intellectual Property Organization's (WIPO) Technology Trends 2021 report developed by 72 leading researchers worldwide:

> For the world of assistive tech, the market is set to expand. Currently, more than 1 billion people globally need at least one assistive product, a figure that is expected to double in the next 10 years as populations age. Even more people will benefit from assistive technology—such as wearables, customized solutions and connected and smart devices—as these technologies spread to the wider population through new consumer goods.
>
> *(WIPO, 2021, p. 7)*

Their collective data indicates that augmentation technologies for physical enhancement are complementing, not replacing conventional assistive technologies. Moreover, one member, Phyllis Heydt from the Office of the WHO Ambassador for Global Strategy, states, "For an individual who receives appropriate assistive technology from childhood their income will increase by an average of USD 100,000 in their lifetime" (p. 21).

Throughout this book, we also acknowledge that mounting rhetoric surrounding assistive or enabling technologies can frame persons with disabilities as a rhetorical motivator used "to bolster the public image of corporations, but at the same time disabled people are negatively depicted as in need of assistive technology or 'fixing'" (Tucker, 2017, p. 30). Fabric metadata includes the keyword *ableism*, which is used by archivists to designate the way that some artifacts could be considered discriminatory within a context. Wherever possible, Fabric archives artifacts by persons with disabilities as creators and contributors (scholars, inventors, artists, entrepreneurs, and/or participants) to voice their own opinions on emerging embodied technologies. One curated collection in Fabric, *Transhumanism, Technology, and Disability: A Critical Perspective* (2020) by Sharon Caldwell and Danielle Rydstedt, presents multiple viewpoints through numerous relevant artifacts.

We also consulted terminology of emerging assistive technology from WIPO's Technology Trends 2021 report (p. 29), which classifies some of the same technologies. We include Figure 2.12 as TPCs are integral to work across the six industries identified in its inner circle, a summary of enabling technologies by the WIPO and a broad way to contribute to defining augmentation, which we reviewed when creating Table 1.1: Scope of Augmentation Technologies. In Figure 2.12, we see the outer circle as representing critical areas for TPC involvement across augmentation technology design, adoption, and adaptation.

Algorithmic control within each assistive device will become much more embodied in the future. Our fitness trackers, smartwatches, phones, and tablets largely constitute our digital selves now, moving with us, monitoring our activity, and staying in constant communication. Pedersen (2020), in her in-depth discussion of the body as a platform, argues that the human drive for seamless, constant connection invariably leads us to the assumption that our bodies will be the interface:

> In a projected future, body networks will connect and report on such things as human thoughts, memories, and feelings, along with organ functionality, biochemistry, and brainwaves. Bodies will participate in cooperative relationships with other human and nonhuman actors and digital infrastructures.
>
> *(p. 25)*

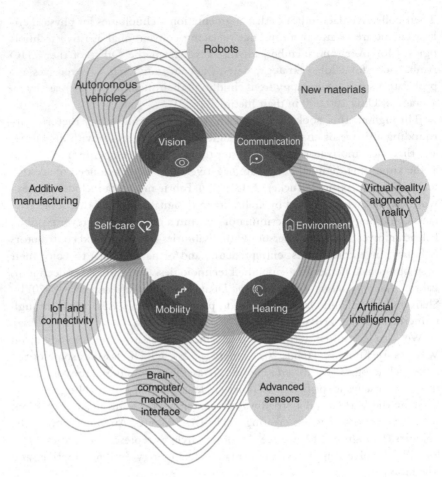

FIGURE 2.12 Enabling technologies from the WIPO Technology Trends 2021 report Figure 1.3. Creative Commons Attribution 3.0 IGO license. https://creativecommons.org/licenses/by/3.0/igo/

She further explains:

> With the coming era of body networks, the proposition is to progress seamlessness one more step, whereby the body is the channel (Astrin et al., 2009). The idea is to make bodily monitoring (like that of skin tech) and biometric data transfer more active and direct.
>
> *(p. 31)*

Along with this speculation is the concept of ambient interfaces that bring the body into a complete ecosystem of "topographical [wearable devices], visceral [implanted sensors], and ambient relationships with the body" (Pedersen & Iliadis, 2020, p. xi). Envisioning the body as a platform brings together cognitive, sensory, emotional, and physical enhancement. For example, in this scenario, platforms currently designed for urban infrastructures, the smart city paradigm, will enfold human subjects in dataspheres. Again, we contend that TPCs must prepare for such embodied algorithmic control that will evolve to interpret further monitoring of cognitive, sensory, emotional, and physical enhancement. Each of these dimensions and exemplary technologies brings new challenges in terms of TPC understanding audience, data, privacy, intellectual property, usability, and content management.

We also acknowledge the rich overlap of many disciplines. In his review of wearable technology research at MNCAT Discovery and EBSCOhost databases, Tham (2018) identified seven dimensions of interactivity for use in "designing robust yet effective user experience in immersive environments": (1) reciprocity/ease of response, (2) synchronicity/context awareness, (3) connectedness/ubiquity/pervasiveness, (4) navigability/accessibility, (5) user control/personalization, (6) entertainment/sensibility, and (7) sensory stimulation/multimodality. In this work he defines interactivity based on "process, features, perception, or the combination of these," noting the need for TPC to "go beyond mouse clicks and screen navigation." He cites work by Cirani and Picone (2015), who "envisioned the interactive characteristics these devices should have, including augmentation of real-life scenarios; integration of social, cyber, and physical worlds; centralization or unification of devices; and proactiveness (automated actions with minimal or no human input)." Tham's dimensions of interactivity support TPC's continued emphasis on working with interface designers on design features that enhance communication, e.g., to post comments to websites or to develop content easily read or accessed on multiple devices; to increase device awareness by creating algorithms for "understanding the rhetorical situation of communication (user, audience, purpose, content, and medium)"; to allow even greater customization for user control over their devices; to foster greater sensory input as a means to activate and navigate device functions; to continue to identify and understand the affordances and limitations of each immersive technology; and even to design devices that are more engaging and entertaining. We strongly agree with the need for TPC to "go beyond mouse clicks and screen navigation," again calling for even greater attention to the future of technical and professional communication amid the dynamic dimensions of augmentation technologies and AI.

References

AHs—Augmented Humans International Conference (2021, January 27). *Wearable reasoner: Enhanced human rationality through a wearable audio device* [Video]. YouTube. https://www.youtube.com/watch?v=9yQ-haXhZIA

AI Med. (2021, February 9). *Looking at a brighter medical future with Google Glass.* Artificial Intelligence in Medicine. https://ai-med.io/analysis/looking-at-a-brighter-medical-future-with-google-glass/

Astrin, A., Li, H.-B., & Kohno, R. (2009). Standardization for body area networks. *IEICE Transactions on Communications, E92.B* (2), 366–372.

Burke, K. (1950). *A rhetoric of motives.* University of California Press.

Caldwell, S., & Rydstedt, D. (2019, November). *Transhumanism, technology, and disability: A critical perspective* [multimedia collection]. Fabric of Digital Life. https://fabricofdigitallife.com/Browse/objects/facets/collection:25

Cesari, V., Galgani, B., Gemignani, A., & Menicucci, D. (2021). Enhancing qualities of consciousness during online learning via multisensory interactions. *Behavioral Sciences, 11*(5), 57. https://doi.org/10.3390/bs11050057

Cirani, S., & Picone, M. (2015). Wearable computing for the internet of things. *IT Professional, 17*(5), 35–41. https://doi.org/10.1109/MITP.2015.89

Crawford, K. (2021). *Atlas of AI: Power, politics and the planetary costs of artificial intelligence.* Yale University Press.

Danry, V., Pataranutaporn, P., Mao, Y., & Maes, P. (2020). *Wearable reasoner: Towards enhanced human rationality through a wearable device with an explainable AI assistant.* Association of Computing Machinery. https://dam-prod.media.mit.edu/x/2020/03/24/AH20_Wearable_Reasoner.pdf

Duin, A. H., & Breuch, L. K. (in press). Writer identity, literacy, and collaboration: 20 technical communication leaders in 2020. In L. Arduser (Ed.), *Workplace writing.* CSU Press.

Duin, A. H., & Pedersen, I. (2023). *Dimensions, scope and classification for augmentation technologies* [Multimedia collection]. Fabric of Digital Life. https://fabricofdigitallife.com/Browse/objects/facets/collection:70

Duin, A. H., Pedersen, I., & Tham, J. (2021). Building digital literacy through exploration and curation of emerging technologies: A networked learning collaborative. In N. B. Dohn, J. J. Hansen, S. B. Hansen, T. Ryberg & M. de Laat (Eds.), *Conceptualizing and innovating education and work with networked learning* (pp. 93–114). Springer. https://doi.org/10.1007/978-3-030-85241-2_6

Emoshape Inc. (2018, May 3). *Emotion chip EPU for real-time emotion synthesis and reasoning* [Video]. YouTube. https://www.youtube.com/watch?v=OUYRBrPG9ZA

Engelbart, D.C. (1962). *Augmenting human intellect: A conceptual framework.* Summary Report AFOSR-3233. Stanford Research Institute, Menlo Park, CA.

Finholt, T. (2002). Collaboratories. *Annual Review of Information Science and Technology, 36*, 73–107. http://www.digitalhumanities.org/companion/

Gartner. (n.d.). *Augmented intelligence.* Gartner Glossary. https://www.gartner.com/en/information-technology/glossary/augmented-intelligence

Ham, B. (2021). *Engineers create a programmable fiber.* MIT News. https://news.mit.edu/2021/programmable-fiber-0603

Hardesty, L. (2018). *Computer system transcribes words users 'speak silently''.* MIT News. https://news.mit.edu/2018/computer-system-transcribes-words-users-speak-silently-0404

Huber, J., Shilkrot, R., Maes, P., & Nanayakkara, S. (Eds.). (2018). *Assistive augmentation.* Springer. https://doi.org/10.1007/978-981-10-6404-3

Hurwitz, J., Morris, H., Sidner, C., & Kirsch, D. (2020). *Augmented intelligence: The business power of human-machine collaboration.* CRC Press. https://doi.org/10.1201/9780429196645

The IEEE Global Initiative on Ethics of Autonomous and Intelligent Systems. (2019). *Ethically-alligned design: A vision for prioritizing human well-being with autonomous and intelligent systems, First Edition.* IEEE Advancing Technology for Humanity. Retrieved from https://standards.ieee.org/content/dam/ieee-standards/standards/web/documents/other/ead1e.pdf

IEEE Digital Reality. (n.d.). *What is augmented intelligence?* IEEE. https://digitalreality. ieee.org/publications/what-is-augmented-intelligence

Iliadis, A., & Pedersen, I. (2018). The fabric of digital life: Uncovering sociotechnical tradeoffs in embodied computing through metadata. *Journal of Information Communication and Ethics in Society, 16*(1). https://doi.org/10.1108/JICES-03-2018-0022

Jensen, S. R., Nagel, S., Brey, P., Ditzel, T., Rodrigues, R., Broadhead, S., & Wright, D. (2018). *SIENNA D3.1: State-of-the-art review: Human enhancement (V1.1).* Zenodo. https://doi.org/10.5281/zenodo.4066557

Jones, B. (2017, August 30). *A bionic pop superstar is going to war for science.* Futurism. https://futurism.com/a-bionic-pop-superstar-is-going-to-war-for-science

Kapur, A. (2018, April 18). *AlterEgo: Interfacing with devices through silent speech.* [Video]. YouTube. https://www.youtube.com/watch?v=RuUSc53Xpeg

Kapur, A. (2019). How AI could become an extension of your mind | Arnav Kapur. *TEDTalks.* https://www.ted.com/talks/arnav_kapur_how_ai_could_become_an_extension_of_your_mind?language=en

Kennedy, K. (2017). Designing for human-machine collaboration: Smart hearing aids as wearable technologies. *Communication Design Quarterly, 5*(4). https://doi.org/10.1145/3188387.3188391

Maes, P. (1995). Agents that reduce work and information overload. In R. Baecker (Ed.), *Readings in human–computer interaction* (pp. 811–821). Elsevier.

Nextech (2021, January 26). *CES 2021: Augmented reality and artificial intelligence steal the show.* Nextech AR. https://www.nextechar.com/blog/augmented-reality-steals-the-show-at-ces-2021

Nichols, G. (June 8 2018). AR for the blind is straight out of Star Trek. *ZDNet.*

Palmer, C. L. (2004). Thematic research collections. In S. Schreibman, R. Siemens, & J. Unsworth (Eds.), *A companion to digital humanities.* Blackwell. http://www.digitalhumanities.org/companion/.

Papadopoulou, A., Berry, J., Knight, T., & Picard, R.W. (2019). Affective sleeve: Wearable materials with haptic action for promoting calmness. In N. Streitz & S. Konomi (Eds.), *Distributed, ambient and pervasive interactions* (pp. 304–319). Springer.

Pedersen, I. (2020). Will the body become a platform? Body networks, datafied bodies, and AI futures. In I. Pedersen & A. Iliadis (Eds.), *Embodied computing: Wearables, implantables, embeddables, ingestibles* (pp. 21–47). MIT Press. https://doi.org/10.7551/mitpress/11564.003.0004

Pedersen, I., & Baarbé, J. (2013). *Archiving the "Fabric of Digital Life".* IEEE International Symposium on Mixed and Augmented Reality - Arts, Media, and Humanities *(ISMAR-AMH)*, pp. 1–6, https://doi.org/10.1109/ISMAR-AMH.2013.6671260

Pedersen, I., & Duin, A. H. (2021). Defining a classification system for augmentation technology in socio-technical terms. *IEEE International Symposium on Technology and Society (ISTAS)* (pp. 1–4). https://doi.org/10.1109/ISTAS52410.2021.9629174

Pedersen, I. & DuPont, Q. (2017). Tracking the telepathic sublime as a phenomenon in a digital humanities archive. *Digital Humanities Quarterly, 11*(4). http://www.digitalhumanities.org/dhq/vol/11/4/000344/000344.html

Pedersen, I., Everrett, T., & Caldwell, S. (2020). The wearable past: Integrating a physical museum collection of wearables into a database of born-digital artifacts. *Digital Studies/Le Champ Numérique*, *10*(1). http://doi.org/10.16995/dscn.366

Pedersen, I., & Iliadis, A. (Eds.). (2020). *Embodied computing: Wearables, implantables, embeddables, ingestibles*. MIT Press.

Pedersen, I., & Mirrlees, T. (2017). Exoskeletons, transhumanism, and culture: Performing superhuman feats. *IEEE Technology & Society Magazine*, *36*(1), 37–45. https://doi.org/10.1109/MTS.2017.2670224

Petit, O., Cheok, A. D., Spence, C., Velasco, C., & Karunanayaka, K. T. (2015). Sensory marketing in light of new technologies. *Proceedings of the 12th International Conference on Advances in Computer Entertainment Technology (ACE '15)*, New York, NY, USA. https://doi.org/10.1145/2832932.2837006

Picard, R. W. (1997). *Affective computing*. MIT Press.

Picard, R. W. (n.d.). *Affective computing*. MIT Press. [Press Release].

Picard, R. W. (2007, December 22). *Affective computing*. Wayback Machine. https://web.archive.org/web/20071222054934/http://www.scholarpedia.org/article/Affective_computing

Project AlterEgo. (n.d.). *Overview*. MIT Media Lab. https://www.media.mit.edu/projects/alterego/overview/

Rodriguez, J. (2019, November 26). *15 examples of human augmentation in everyday life*. Fresh Consulting. https://www.freshconsulting.com/insights/blog/examples-of-human-augmentation/

Sprenger, R. (2018, June 26). Beyond bionics: how the future of prosthetics is redefining humanity. *Guardian*. https://www.theguardian.com/world/video/2018/jun/26/beyond-bionics-how-the-future-of-prosthetics-is-redefining-humanity-video

Starner, T. E. (2002, July-September). The role of speech input in wearable computing. *IEEE Pervasive Computing*, *1*(3), 89–93. https://doi.org/10.1109/MPRV.2002.1037727

Tham, J. (2018). Interactivity in an age of immersive media: Seven dimensions for wearable technology, Internet of Things, and technical communication. *Technical Communication*, *65*(1), 45–65.

Thilmany, J. (2019, February 27). *Exoskeletons for construction workers are marching onsite*. Constructible. https://constructible.trimble.com/construction-industry/exoskeletons-for-construction-workers-are-marching-on-site

Time. (2021, September 22). *Viktoria Modesta & Katie McIntyre & Nina Hawkins*. https://time.com/collection/timepieces-nft/6098981/viktoria-modesta/

Tucker, B. (2017, April 26). Technocapitalist disability rhetoric: When technology is confused with social justice. *Enculturation: A Journal of Rhetoric, Writing, and Culture*. http://enculturation.net/technocapitalist-disability-rhetoric

Upskill. (n.d.). *Upskill and Boeing augmented reality: Reinventing aerospace manufacturing and supply chain operations*. https://upskill.io/landing/upskill-and-boeing/

Valeriani, D., Cinel, C., & Poli, R. (2019). Hybrid collaborative brain–computer interfaces to augment group decision-making. In H. Ayaz & F. Dehais (Eds.), *Neuroergonomics* (pp. 187–190). Academic Press. https://doi.org/10.1016/B978-0-12-811926-6.00031-2

Van Dijk, T. A. (2001). Critical discourse analysis. In D. Tannen, D. Schiffrin, & H. Hamilton (Eds.), *Handbook of discourse analysis* (pp. 352–371). Oxford University Press.

Vanguard Industries Inc. (2022, February 18). *Moflin: An AI pet robot with emotional capabilities*. Kickstarter. https://www.kickstarter.com/projects/vanguardindustries/moflin-an-ai-pet-robot-with-emotional-capabilities

Velasco, C., Obrist, M., Petit, O., & Spence, C. (2018). Multisensory technology for flavor augmentation: A mini review. *Frontiers in Psychology, 9*(26). https://doi.org/10.3389/fpsyg.2018.00026

Vergun, D. (2017, June 1). *Heads-up display to give soldiers improved situational awareness*. US Army. https://www.army.mil/article/188088?a

Vi, C. T., Gatti, E., Ablart, D., Velasco, C., & Obrist, M. (2017). Not just see, but feel, smell, and taste the art: A case study on the creation and evaluation of multisensory art experiences in the museum. *International Journal of Human-Computer Studies, 108*, 1–14. http://sro.sussex.ac.uk/id/eprint/68811/17/1-s2.0-S1071581917300988-main%20(1).pdf

Wigmore, I. (2019, February). *Augmented intelligence*. Whatis.com. https://whatis.techtarget.com/definition/augmented-intelligence

WIPO (2021). *WIPO technology trends 2021: Assistive technology*. World Intellectual Property Organization. https://www.wipo.int/edocs/pubdocs/en/wipo_pub_1055_2021.pdf

3
AGENCY, AFFORDANCES, AND ENCULTURATION OF AUGMENTATION TECHNOLOGIES

Overview

Augmentation technologies are undergoing a process of enculturation due to many factors, one being the rise of artificial intelligence (AI), or what the World Intellectual Property Organization (WIPO) terms the "AI wave" or "AI boom." Chapter 3 focuses critical attention on the hyped assumption that sophisticated, emergent, and embodied augmentation technologies will improve lives, literacy, cultures, arts, economies, and social contexts. The chapter begins by discussing the problem of ambiguity with AI terminology, which it aids with a description of the WIPO Categorization of AI Technologies Scheme. It then draws on media and communication studies to explore concepts such as agents, agency, power, and agentive relationships between humans and robots. The chapter focuses on the development of non-human agents in industry as a critical factor in the rise of augmentation technologies. It looks at how marketing communication enculturates future users to adopt and adapt to the technology. Scholars are charting the significant ways that people are drawn further into commercial digital landscapes, such as the Metaverse concept, in post-internet society. It concludes by examining recent claims concerning the Metaverse and augmented reality.

Key Questions

1 How is AI emerging to generate augmentation technologies?
2 Amid emerging augmentation technologies, how are humans and AI technologies framed as agents with agency?

DOI: 10.4324/9781003288008-4

3   How is the massive corporate backing of Metaverse as the next phase of the internet changing the design, adoption and adaptation of augmentation technologies?
4   How are inventors and stakeholders envisioning cognitive, sensory, emotional, and physical enhancement?

Chapter 3 Links

Throughout the chapter, we refer to articles, videos, and reports which can be found in a related chapter collection at Fabric of Digital Life called *Agency, affordances, and enculturation of augmentation technologies* (Duin & Pedersen, 2023). You can find a link to this collection at https://fabricofdigitallife.com.

Introduction

Susan B. Anthony declared in 1896 that bicycling had "done more for the emancipation of women than anything else in the world" (Bly, *New York World*, p. 10). Sarah Hallenbeck (2012) begins her article on user agency in technical communication with this quote, noting how Anthony's statement was:

> characteristic of the cultural moment in which she was writing—in which a bicycle craze had infected people in nearly every class, and the demand for 'the steed which never tires' was so high… that many bicycle factories operated around the clock,

with women celebrating the transformative power of this new technology that increased "mobility, independence, and physical health and strength" (p. 290). However, "moral conservatives warned that bicycling, as well as the changes in attire that it required, would 'unsex' women riders" (p. 291). Despite these warnings, women's riding only increased, and bicycle manufacturers began to add features in support of "the embodied practices of women themselves," constituting a "regendering of the bicycle, whereby the once masculine object now developed strong associations with female riders" (p. 291). Hallenbeck investigates how these early women riders, "as users excluded from official avenues of design and manufacture" (p. 291), laid claim to this technology not originally designed or intended for their use. She complicates the notion that technology "becomes a stable and static object once it enters into common use" (p. 291). To do so, she extends her focus to contexts of use, highlighting what sociologist Andrew Pickering (1995) called a "dance of agency," noting that in this configuration, "users are not merely passive recipients of producers' expert knowledge and inventive genius but also shapers of the cultural and symbolic meanings attached to the technologies with which they interact" (Hallenbeck, p. 291, citing Oudshoorn & Pinch, 2003).

More recently, Rosi Braidotti (2021) contends that "the posthuman turn can result in a renewal of subjectivities and practices by situating feminist analyses productively in the present" (p. 106). Over the past decade, TPC scholars have embraced the social justice turn by "recognizing the injustices and oppressive systems embedded in our work as technical communicators" (p. 133). Rebecca Walton states in an interview that "intersectionality is an important part of our thinking right now because it attunes us to the compounding nature of oppressions for people who are multiply marginalized" (Walton et al., 2019, p. xxiii). An ethical framework for augmentation technologies needs to revise older patterns of thinking and acting in the field. Walton et al. (2019) contend that "Social injustices require coalitional action, collective thinking, and a commitment to understanding difference that is not necessarily demanded by other technical communication problems" (p. 1).

Complicating our field is the fact that we are in the midst of an augmentation technology craze fueled partly by a blind trust in AI. The WIPO names the phenomenon an "AI wave" (p. 19), calling it a "good time to take a close look at the state of research and exploitation of AI technologies" (p. 20). Bernard Marr (2020), writing for *Forbes*, identifies AI as:

> undoubtedly one of the biggest tech trends at the moment, and during 2021 it will become an even more valuable tool for helping us to interpret and understand the world around us. The volume of data we are collecting on healthcare, infection rates, and the success of measures we take to prevent the spread of infection will continue to increase. This means that machine learning algorithms will become better informed and increasingly sophisticated in the solutions they uncover for us.

He continues by citing robotics as a means to reduce human labor costs and balance uncertainty and customer demand, and citing extended reality (XR), virtual reality (VR), and augmented reality (AR) as tools to help us avoid potentially dangerous situations, e.g., allowing medical examinations and diagnosis to be carried out remotely, providing real-time warnings about virus spread, and guiding us through the challenging world that we face. Likewise, Brian Chen (2021), writing for *The New York Times*, highlights our embrace of augmentation technologies to interact with others and with physical worlds now less safe given the pandemic. And XPRIZE stories showcase humans breaking boundaries with myriad uses of augmentation technology; for example, novelist Kazuo Ishiguro says:

> When you get to the point where you can say that [a] person is actually intellectually or physically superior to another person because you

have removed certain possibilities for that person getting ill… or because they're enhanced in other ways, that has enormous implications for very basic values that we have.

(XPRIZE, 2021)

AI Technologies and Enculturation

In previous chapters, we have explained that the rise of AI, in and of itself, generates the rise of augmentation technologies. We argue that augmentation technologies are undergoing a process of normalization due to massive transformations in AI leading to the enculturation of both. However, definitions for AI are ambiguous. One of the issues TPC professionals are facing is the lack of clarity about what AI means:

AI is "the ability of certain machines to do things that people are inclined to call intelligent."
Interview with Marvin Minsky in *The New Yorker.*

(Bernstein, 1981)

AI can be defined as "Systems that think like humans, systems that act like humans, systems that think rationally, systems that act rationally."
Stuart Rusell and Peter Norvig in *Artificial Intelligence: A Modern Approach.*

(Russell & Norvig, 1995)

AI is "the 60-year-old quest to make machines capable of mental or physical tasks seen as emblematic of human or animal intelligence."
Tim Simonite, *Wired* Magazine.

(Simonite, 2019)

Artificial general intelligence (AGI) can be defined as "highly autonomous systems that outperform humans at most economically valuable work."

OpenAI Charter (OpenAI, 2018)

To deal with shifting concepts in this chapter, we categorize AI technology under three terms for the sake of discussion in this book. The terminology is drawn from the Association for Computing Machinery (ACM) Computing Classification Scheme, as quoted by the WIPO (2019). WIPO explains the intent for better clarity: "this scheme has the advantage of providing a clear analytical framework for the report and the presentation of the evolution of AI technologies over time" (p. 25).

Figure 3.1 illustrates the category called *AI techniques*, which are "advanced forms of statistical and mathematical models, such as machine learning, fuzzy

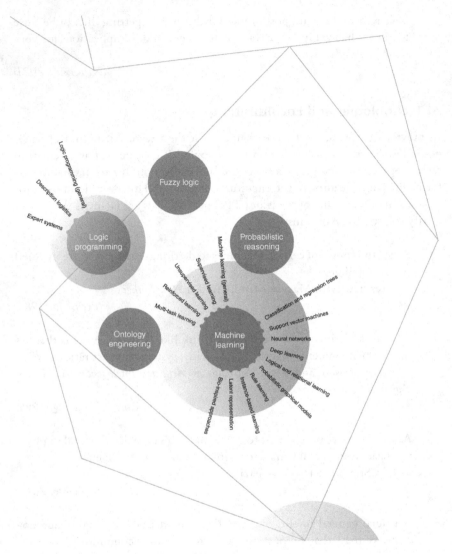

FIGURE 3.1 AI techniques (WIPO, 2019, p. 24). Creative Commons Attribution 3.0 IGO license. https://creativecommons.org/licenses/by/3.0/igo/.

logic and expert systems, allowing the computation of tasks typically performed by humans" (p. 25). AI techniques should be seen as the technical means to create *AI functions* within applications.

Figure 3.2 illustrates *AI functional applications*, which include "functions such as speech [processing, predictive analytics] or computer vision which can be realized using one or more AI techniques" (p. 25). Computer vision informs and enhances *augmented reality*, which is already a developed computing

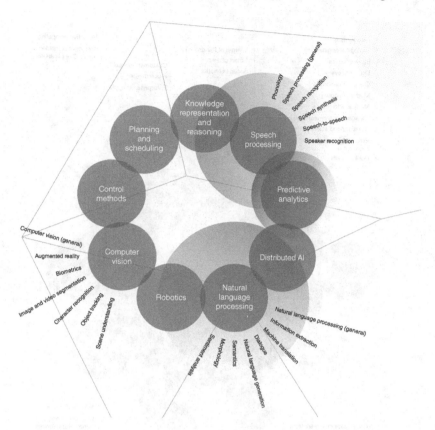

FIGURE 3.2 AI functional applications (WIPO 2019, p. 26). Creative Commons Attribution 3.0 IGO license. https://creativecommons.org/licenses/by/3.0/igo/.

field. But this infusion of AI techniques makes AR much more sophisticated than before.

Figure 3.3 illustrates *AI application fields*, which include the many "fields, areas or disciplines where AI techniques or functional applications may find application, such as transportation, agriculture or life and medical sciences" (p. 25).

Words describing AI often appear as common buzzwords in popular technology discourses. But one can better imagine AI technologies in the stratified WIPO scheme whereby concepts such as *smart cities* are constituted through a hierarchy of the categories (see Figure 3.4). For example, *AI techniques* can be combined to create *AI functional applications,* which lead to fields like *Networks.* Networks help to constitute Internet of Things (IoT), smart cities, and social networks in working structures where humans interact with interfaces. Figure 3.4 illustrates the scheme.

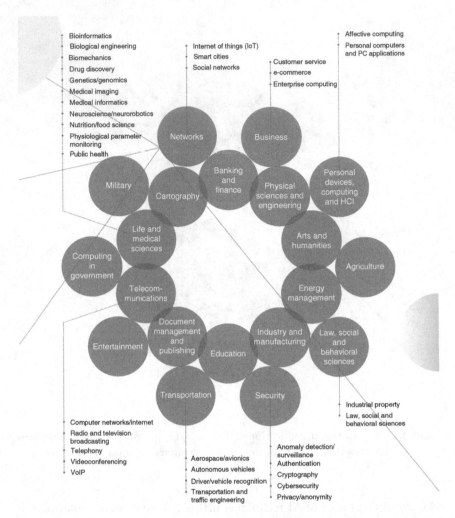

Bioinformatics
Biological engineering
Biomechanics
Drug discovery
Genetics/genomics
Medical imaging
Medical informatics
Neuroscience/neurorobotics
Nutrition/food science
Physiological parameter monitoring
Public health

Internet of things (IoT)
Smart cities
Social networks

Customer service
e-commerce
Enterprise computing

Affective computing
Personal computers and PC applications

Networks
Business
Banking and finance
Military
Cartography
Physical sciences and engineering
Personal devices, computing and HCI
Life and medical sciences
Arts and humanities
Computing in government
Agriculture
Telecommunications
Energy management
Document management and publishing
Entertainment
Industry and manufacturing
Education
Law, social and behavioral sciences
Transportation
Security

Computer networks/internet
Radio and television broadcasting
Telephony
Videoconferencing
VoIP

Aerospace/avionics
Autonomous vehicles
Driver/vehicle recognition
Transportation and traffic engineering

Anomaly detection/surveillance
Authentication
Cryptography
Cybersecurity
Privacy/anonymity

Industrial property
Law, social and behavioral sciences

FIGURE 3.3 AI application fields (WIPO 2019, p. 27). Creative Commons Attribution 3.0 IGO license. https://creativecommons.org/licenses/by/3.0/igo/.

Agency in the Digital Age amid Human and Machine Cultures

This book starts with a commitment to reconceptualize augmentation technologies through a human-centric, ethical framework, acknowledging the rich assemblages within which interactions occur between humans and non-human agents. In the first chapter, we pointed to the hyped rhetorics of betterment that often accompany these enhancement technologies, which sometimes obscure our ability to assess their use. Taking it a step further, Zizi Papacharissi (2019) writes of the theme of augmentics, "if we imagine and then design technology as something that augments or enhances our human powers, then we are bound

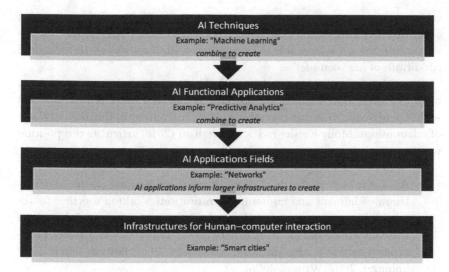

FIGURE 3.4   Explanation of WIPO categorization of AI technologies scheme.

to wind up with bizarre superhero complexes, to say the least, and grossly underestimate *techne*" (p. 7). She does, however, move toward a more ameliorative vision for augmentics. She writes, "we have the opportunity to become different, and with this opportunity comes another possibility, perhaps an obligation: to reimagine how we do things; to do things differently; to not fall into the trap of reproducing what we already have" (p. 7).

In communication and media studies as well as technical communication theory, agency is associated with people, with users. To use a bicycle, an agent (the rider) must have the agency or capacity to do so. According to Kris Cohen (2022), "expressed simply, an agent is one who acts. The power granted or effected through that action is the quality of agency.... Any definition of agency is complicated from the start by disciplinary differences in its conceptualization and use." AI, sociology, and history disciplines cite "social agents" as empirical or quasi-empirical entities; political and feminist theories view agency as a tool aligned with subjectivity and marginalization; and philosophical work on autonomy sees agency as a relation of self to self. According to Michel Foucault (1982), agency is a concept based on power relations; for Martin Heidegger (1977), agency is defined within new technologies of representation, e.g., within the "collective" of the agent (set of scientists, observers); and for Marshall McLuhan (1964), technology acts on human agents to extend perception and cognition, with the clear danger of being blinded to the ways technologies impact our existence. Karlyn Kohrs Campbell (2005) defines agency as "polysemic and ambiguous, a term that can refer to invention, strategies, authorship, institutional power, identity, subjectivity, practices, and subject positions" (p. 1). Cohen (2022) states that all of this literature denotes the effective agent as "the powerful self, the assertive self, the self who is able to impose her will, if only over the terms and conditions of her own life."

Bruno Latour's (1999) Actor-Network Theory (ANT) denotes agency as a matter of attribution; i.e., that things have agency as a result of their position within a network, and that everything exists in constantly shifting networks of relationships. Molly Kessler and Scott Graham (2018) articulate this position within technical communication:

> ANT has been primarily used in efforts to understand (1) the coarticulation of humans and nonhumans institutions working together to accomplish a goal or set of goals, and (2) the complexity and density of these networks, and (3) the nature of power and agency as network properties (Graham & Herndl, 2011; Miettinen, 1999; Potts & Jones, 2011; Spinuzzi, 2007; Winsor, 2006).
>
> *(p. 123)*

ANT scholarship through these recent decades has deepened our understanding of the social lives of technologies, elucidating how agency is embedded in a broader sociocultural landscape (Neff & Nagy, 2019), with specific studies such as Kessler and Graham's (2018) examination of the embedding of agency in prescription drug labels.

Advancements in recent technologies challenge these fundamental notions of agency, as "tools and techniques including machine learning, AI, and chatbots may be capable of exercising complex 'agentic' behaviors" (Neff & Nagy, 2019, p. 97). As discussed in Chapter 1, robotic companions now read human emotions, and chatbots evoke reactions to help people make decisions or cope with anxiety. In Chapter 2, we classified this interactivity as emotional enhancement. Calling for a better definition of agency amid these "complex, technologically mediated interactions," Neff and Nagy ask, "Does the agency of the people and, increasingly, of things that we choose to communicate with matter?" (p. 97). They employ Albert Bandura's (2006) social-cognitive psychological theory of human agency to "reframe and redefine agency in terms of how and when people interact with complex technological systems.... Being an agent means that people can exert intentional influence over their mental processes and physical actions to achieve their desired outcomes" (p. 98). In short, people delegate tasks to augmentation technologies as their personal assistants and companions, resulting in a symbiotic relationship. Using Bandura's theory, Neff and Nagy link agency to intentionality, forethought, and self-reflectiveness. Collective agency indicates shared beliefs and goals, and proxy agency involves other people or tools that can help one achieve goals. They propose the term "symbiotic agency" as a specific form of proxy agency, i.e., when people interact with technologies:

> They exercise proxy agency through a technologically mediated mode referring to the entanglement of human and non-human agencies. For symbiosis, similar to entanglement, implies an obligated relationship— symbionts are completely dependent on each other for survival, it can be considered a proxy agentic relationship that may provide different benefits for the recipients.
>
> *(p. 102)*

People can benefit from, not be affected by, or be harmed by symbiotic relationships. They note how the "Uncanny Valley Effect" (Reichardt, 1978):

> documents the discomfort that people feel when robotic agents are seen as having a too humanlike appearance.... When ascribing agency to complex technologies, such as chatbots, robots, or other AI applications, people project their emotions to these entities and try to find explanations for their 'behaviors' (Darling, 2014)
>
> *(p. 103)*

Neff and Nagy explain that these relationships can also be parasitic rather than symbiotic, noting "users may also feel that they are increasingly getting 'addicted to' or 'controlled by' technologies due to the perceived parasitic attributes of these tools." (p. 103).

We contend that augmentation technologies elicit symbiotic agency; any lines between what users do and what technologies do are now forever blurred. As augmentation technologies listen in on conversations and read thoughts, they learn and shape future meaning and responses, resulting in increasingly intimate relationships. "By redefining human–technology interaction as a unique form of proxy agency, [we] can investigate both the human agentic properties and the technological attributes of what it means to be a networked self" (Neff & Nagy, 2019, p. 104).

In short, computers adapt to humans. The affordances for directing computers include speech, gestures, eye gaze, and even human electrophysiological signals (Raisamo et al., 2019). We gain direction while driving that includes auditory as well as haptic smartwatch notification to make sure to turn at the appropriate time. Roope Raisamo et al. discuss such information flow and associated technologies in wearable augmentation: "Numerous sensors and cloud data provide information, artificial intelligence filters it and it is presented in easy-to-understand ways to support human cognition in a timely manner. Physical tools or robots enable action in and changes to the environment" (p. 136). Moving forward, technological ability to sense humans will intensify as the

Internet of the Body evolves. Technologists are orienting development toward human-centric values: "only a system that is able to extend its capabilities to perceive and adapt to human actions, intentions, and emotional states, can be considered a truly smart system" (Fernandes et al., 2022).

In effect, we have created human and machine cultures, the posthuman complete with cognitive assemblage, technical agency, and human interactions. N. Katherine Hayles (2002), prominent posthuman theorist, states that "Agency still exists, but for the posthuman it becomes a distributed function" (p. 319). She clarifies our lived experience as follows: "Living in a technologically engineered and information-rich environment brings with it associated shifts in habits, postures, enactments, perceptions—in short, changes in the experiences that constitute the dynamic lifeworld we inhabit as embodied creatures" (p. 299). Reaffirming her position in a later book (Hayles, 2017), she explains post-digital human and non-human collaboration in terms of "complex human technical assemblages in which cognition and decision making powers are distributed throughout the system. I call the latter cognitive assemblages" (p. 4).

Non-human Agents Become Enculturated in Industry Sectors

How are non-human agents emerging in industry sectors in terms of collective agency? How is ethical alignment or shared beliefs and goals appearing in the discourse of emergence, if at all? In a recent exploratory study to identify the nature of emergent relationships between human and non-human agents, we analyzed how working assemblages of activity are taking shape. The concentration was on AI technologies marketed for learners in networked learning environments because we wanted to better understand how AI is proposed to augment humans in this context (Pedersen & Duin, 2022). We looked at corporate advertisements for new educational technologies, as well as ones that simply claimed to have an educational function. Several were still in corporate pre-release research and development hyping future products to help raise funds and awareness for startups. We did not include university or government research because the goal for this study was to target products that will enter classrooms and people's homes, ones that are closer to release. In Chapter 2, we discussed claims about augmentation that drive adoption and adaptation. Our intent with this study was informed by the same research goal, to reveal both affordances and consequences of these relationships. We used Fabric of Digital Life (https://fabricofdigitallife.com/) metadata to identify a dataset of 17 video advertisements for AI inventions. It covered wearables, mobile, and robotic devices. Marketing materials also included sales copy about AI and how an AI agent, for example, could improve human skills. We applied a critical discourse analysis framework to reveal these persuasive statements in the advertising. We analyzed the way that children were visually depicted with robots and human teachers in order to understand how human agents and AI agents are promoted in collaborative cognitive assemblages. At the same time, robot actions and behaviors (signifiers) are realized through composites of sensors, trackers, cameras, moving mechanical parts, and AI technologies, etc., built to operate as if a unified agent. This information can be tracked through the technology keyword category. Fabric provides grounds for analysis through the metadata. Styles of communication change with each new technical advancement signaling the continuous possibility of novel augmentation.

The advertisements also frame agentive relationships through values expressed to promote their adoption. A good example is Little Sophia by Hanson Robotics, a spin-off of the famous Sophia robot (see Figure 3.5). Little Sophia is a consumer-grade device still in pre-release with the potential to be used in homes and classrooms. The suggested scenarios for Little Sophia appear in the ad copy: "with Little Sophia's software, and included tutorials through Hanson's AI Academy, she is a unique programmable, educational companion for kids, inspiring children to learn through a safe, interactive, human-robot experience" (Hanson Robotics, 2019). Little Sophia is also always enveloped in Sophia's fame as a much more developed AI agent/celebrity that takes every

FIGURE 3.5 Sophia at the AI for Good Global Summit 2018. Photography by ITU pictures from Geneva, Switzerland. Creative Commons Attribution 2.0 Generic license. https://creativecommons.org/licenses/by/2.0/

opportunity to promote her value. It is unlikely that any household will have a Sophia robot for decades. It is much more likely that students, workers, and consumers will be involved in adapting to functionality for robots on the scale of Little Sophia that will eventually progress to that of Sophia.

Overall, the study found imbalances in proposed human–non–human relationships:

> that [the] advertising promotes corporate products while also promoting idealized social practices for human–computer interaction and human–robot interaction in learning contexts. Using AI to automate relationships between students and teachers frames AI systems as authorities in both robot and non-robot platforms, blurring and minimizing student and instructor agency in learning environments.
>
> *(Pedersen & Duin, 2022, p. 1)*

We prepared Table 3.1, Rhetorical Propositions about AI and Augmentation Technology in Marketing Videos, to exemplify some of the marketing claims by corporate rhetors. Column 1 identifies the type of augmentation hardware platform about which claims are made. Column 2, "Video artifacts and main augmentation," lists each video and the central augmentation it promises to provide, many of which are abstract value propositions. Column 3, "Persuasive claims about augmentation made by companies," includes quotes from the videos or textual ad copy surrounding the artifacts. Column 4 lists technological keywords used to substantiate those claims.

Each row provides a snapshot of idealized scenarios for an augmentation aligned with the technologies proposed to fulfill them. Each artifact serves to enculturate consumers to adopt a product and adapt to it. For example, Waverly Labs (2019) promises to be working toward "A world without language barriers," a significant augmentation, and we report in column 4 exactly which technologies are being offered to fulfill such a claim. Composites of embodied devices (wearables: earbuds, microphones), AI technologies (neural networks, speech recognition), and internet infrastructure (cloud computing, wireless connectivity) are implicated in fulfilling these rhetorical claims.

TPC practitioners need greater awareness of these components as they rapidly emerge to support their functional use in homes, schools, and/or work scenarios. Robots operating in ambient spaces around humans form cognitive assemblages and increased emotional enhancements. For example, many of the robots use pre-built facial expressions presumably to establish empathetic communication channels by mimicking human body language. TPC will be involved in writing these scripts for robots to legitimize the messages and constitute robots as agents. However, awareness of the affordances of these developing technologies is and will be key (for additional discussion, see Ayanoglu and Duarte, 2019).

TABLE 3.1 Rhetorical propositions about AI and augmentation technology in marketing videos

Hardware Platform	Video artifacts and main augmentation	Persuasive claims about augmentation made by companies	Technological keywords
Wearables			
	Ambassador by Waverly Labs Company: Waverly Labs, 2019 augmentation: Learn new languages	"imagine being able to snap your fingers and become fluent in 20 languages" "A world without language barriers"	Artificial Intelligence (AI), Neural Networks, Speech Recognition, Translators, Wearable Translators, Smart Earbuds, Microphones, Speakers, Real-Time, Cloud Computing, Smartphone Applications, Wireless Connectivity
	FOVE VR Headset: Tracks Subtle Eye Movements in Virtual Reality Company: Fove, Inc., 2015 augmentation: Interact in virtual worlds	"We are on a mission to unlock the essence of reality in virtual worlds" "Enables new forms of expressions, communication and movement"	Virtual Reality (VR), Virtual Worlds, Assistive Technologies, AI, Headsets, Head Mounted Displays (HMD), Cameras, Eye Tracking, Infrared Tracking Systems, High Resolution Display, Natural User Interface (NUI)
Robots			
	Little Sophia by Hanson Robotics Company: Hanson Robotics, 2019 augmentation: Inspire imagination and learning	"She can take you on wild adventures"	Robotics, Humanoid Robots, Social Robots, Human-Robot Interaction (HRI), AI, Facial Recognition, Pre-Built Facial Expressions, Cameras, Microphones, Speakers, Smartphone Applications, Open Source, Wireless Connectivity
	Xiaoyou robots made by Canbot Company: Canbot, 2018 augmentation: Automate teaching, learning, and living	"Robot for your better life"	Robotics, Humanoid Robots, Social Robots, HRI, AI, Natural Language Processing (NLP), Voice Recognition, Intelligent Home Control Systems, Smart Homes, Internet Of Things (IoT), Tablets, Cameras, Microphones, Speakers, Video Calling, Sensors, Obstacle Detection, Remote Monitoring, Wireless Connectivity

(Continued)

Hardware Platform	Video artifacts and main augmentation	Persuasive claims about augmentation made by companies	Technological keywords
Mobile			
	Meet AI XPRIZE Semifinalist Iris. AI Company: Iris.AI, 2020 Augmentation: Keep up with research, become efficient, overcome information overload	"Iris is an AI science assistant to help researchers do their work"	AI, Algorithms, Machine Learning, AI Research Assistants, Text Generators, Text Summarizer, Data Collection, Data Extraction
	MondlyAR—learn languages in AR Company: ATi Studios, Mondly Languages, 2018 Augmentation: Learn new languages	"The app creates a bridge between two worlds: your environment and a virtual universe" "your personal language learning assistant"	AR, Mobile Augmented Reality, AI, Speech Recognition, Virtual Assistants, Virtual Worlds, Cameras, Microphones, Speakers, Smartphone Applications

People project agency *upon* non-human agents. Consider the Amelia Integrated Platform (2022), through which conversational AI/Amelia, a non-human agent, takes on the role of "digital employee" to deliver "the best elements of human interaction—conversation, expression, emotion and understanding—to user experiences every day, driving deeper connections and greater business value." Amelia's agency, both rhetorically and ethically provocative/hyped/problematic, inflates her role as an imagined employee. A recent article describing 50 robots that now work as cleaners, but also as surgeons and physiotherapists, at Singapore's high-tech hospital also refers to these as "AI-powered employees" (Cairns & Tham, 2021). As technical and professional communicators, our task is to understand this evolution, positioning ourselves in collaboration (human-autonomy teaming), with focus on human-centered design. We further discuss digital employees in Chapter 7.

Metaverse, Augmentation Technology, and Virtual Worlds

Corporate playbooks contain initiatives to drive higher revenues by changing the way people use their devices. Many writers are pointing out how people are drawn further into commercial digital landscapes in post-internet society (Mosco, 2017; Andrejevic, 2020; Crawford, 2021). In Chapter 2, we discussed justifications for augmentation, including the goal to enhance the senses to construct an enhanced reality. One dramatic pitch by several "big tech" companies is the Metaverse trope. Definitions for the term "Metaverse" have not solidified. One article names it "a parallel digital reality in which users play and work—and can buy and sell in cryptocurrencies" ("Big Tech's Supersized Ambitions," 2022). Another calls it an "emerging virtual market which could, depending on whom you ask, ultimately generate revenues of between $1trn and $30trn" (MacManus, 2022). More often it is pondered about:

> Will it be an all-consuming futuristic world of virtual reality, avatars, oceanside mansions and other online razzmatazz that will make the real world a dull place by comparison? Or will it simply be a richer, more immersive version of what already exists today?
>
> *("Schumpeter: Lords of the Metaverse," 2021)*

In October 2021, the Metaverse became wildly popular when "Mark Zuckerberg renamed Facebook *Meta* and described humankind's new future in virtual worlds" ("Big Tech's Supersized Ambitions," 2022). In some ways, the Metaverse is an incredibly hyped research project. One article reports that "somewhere between 5% and 20% of the tech giant's massive R&D spending goes towards what, for the purposes of this article, we are calling 'frontier technologies': the metaverse, autonomous vehicles, health care, space, robotics, fintech, crypto and quantum computing" ("Big Tech's Private Passions," 2022). Yet, the Metaverse does not exist in any real capacity; it is a dream: "The metaverse is for the foreseeable future quite literally science fiction: a fully immersive, persistent virtual world where, with the help of high-tech goggles and other kit, people interact, work and play via online avatars of their real-world selves" ("Building a Metaverse with Chinese Characteristics," 2022). Significant sensory enhancements would need to be adopted in order to achieve such a virtual world. Augmentation technologies would provide the means to traverse these interconnected virtual worlds, amounting to the next generation 3D internet.

As one example, watch Joanna Stern, Senior Personal Technology Columnist for *The Wall Street Journal*, in her video *Trapped in the Metaverse: Here's What 24 Hours in VR Feels Like*. She confronts the fantastical claims made about the medium with realistic questions such as "Will we sleep in VR?" "Will we have multiple avatars?" and "How will we navigate between working and socializing in metaverse virtual worlds?" The video gives TPCs a snapshot of usability affordances as well as drawbacks when people are asked to use a VR headset for whole days. It illustrates how multiple human–computer interaction styles are evolving to help people communicate with each other and their own digital information.

In response to the hype, augmentation technology companies have entered the fray, using the Metaverse as a persuasive concept to excite, enculturate, and innovate. Metaverse terminology helps simplify complex technology descriptions, and works to sell people on the idea of enhancement as an alluring and realizable future.

Moreover, the Metaverse concept taps longstanding Big Tech corporate ambitions. Tracing a few of them explains its trajectory and how it will contextualize augmentation technology in future through the themes that are dominating this call for adoption. On July 20, 2009, *New York Times* writer Richard MacManus made one of the many early predictions of a Metaverse in his article, "The Wearable Internet Will Blow Mobile Phones Away." He explained why people should eschew clunky mobile phones and adopt wearable components:

> The Web as we know it today is full of manual steps, such as visiting websites and searching for information. In 10 years time we'd hope that the Web of Data would be much better realized.... You can see the power of this as a next generation Internet interface, as it removes several manual steps from the process of receiving relevant, contextual information about something or someone.... There are also sensors in the objects—for example the book has a barcode that, in combination with the wearable device, will pull down data from Amazon.com via the Web.... [We] are very excited about next-generation Internet interfaces, such as augmented reality and so-called cross reality. These wearable devices strike me as being the most impressive future Web interface that I've seen in a while.

MacManus's 2009 call to design this future internet hangs on interoperability, the desire for more efficiency when interacting with devices and fewer "manual steps" (MacManus, 2009). Now more than a decade later, a "Web of Data" has emerged, and users are very much embedded in algorithmic cultures "receiving relevant, contextual information about something or someone." MacManus also predicted an IoT with the idea of "sensors in the objects" leading to more integration with corporate spheres, such as Amazon. Interestingly, Amazon was worth $59.72 billion in market cap in 2009; now it is worth $1,598 trillion in 2022, justifying its vision (Market Capitalization of Amazon, 2022). Amazon wildly exceeded expectations as an internet and AI company. In 2009, MacManus also plugged immersive media (sensory enhancement) such as "augmented reality" to provide the means to interact more efficiently with his envisioned internet. He argued for more efficiency, automation, and embodiment for internet interactivity, which aligns with our arguments for this book concerning augmentation technology emergence and its promotion.

Fast forward to 2022, and Richard MacManus now writes about the Metaverse's potential for creative content collaboration (MacManus, 2022). He describes "Meta's vision of an interoperable set of competing virtual worlds," but he also describes Nvidia's "Omniverse [which] has officially turned into a content creation platform for virtual worlds on the consumer internet. Nvidia is welcoming 'millions of individual creators and artists' onto its platform, to build for the next generation internet" (MacManus, 2022). Rather than HTML, content creators will write Universal Scene Description (USD); all of Omniverse is built on top of Pixar's open source Universal Scene Description (Pixar, 2022), which Nvidia has pitched as the "HTML of 3D worlds" (MacManus, 2022).

So, how will metaverses be built? Technical and professional communicators can read about USD and shared virtual worlds. Michael Kass, Senior Distinguished Engineer at Nvidia Omniverse, describes USD and writes, "at the core of any true 3D web or the open Metaverse, we need an open, powerful, flexible and efficient way to describe the shared virtual world" (Kass, 2021). *Computer Graphics World* explains how Pixar developed USD over the past 25 years, stating "we have needed ways to describe the 3D scenes we are synthesizing in a way that is mathematical enough for computers to understand, but which is also understandable and manipulatable by technologists *and* artists" (Grassia & ElKoura, 2020).

Metaverse and Augmented Reality

Inventors and stakeholders are envisioning cognitive, sensory, emotional, and physical enhancement for the same reasons they did a decade ago, albeit with more intensity than ever before. The 2022 CES show, already a venue for hyped announcements, featured several augmentation technologies under the Metaverse banner. InWith Corporation claims to provide "potentially the most advanced platform for viewing the coming Metaverse." Rather than announce yet another phone-based application, InWith Corporation explains that the platform would be hosted on the user's eyeballs: "an electronic soft contact lens platform designed for the masses to wear comfortably, enabling an easy transition from real-world to Metaverse, at will" (InWith at CES 2022, 2022). In one broadcast by CNET, the concept is broken down in simple terms promoting the idea that users will be able to flip from real-world interaction (AR) to virtual immersion (VR) much more seamlessly by wearing contact lenses, leading to significant sensory enhancement (Altman, 2022).

We have defined augmentation technologies as those that enhance human capabilities or productivity by adding to the body in the name of efficiency and automation. InWith Corporation soft contacts and platform are offered as an efficient means to access the Metaverse (a concept that does not yet exist in structural terms as we have discussed). Most portraits of Metaverse experiences promote the duality of AR and VR that users will glide into and out of virtual worlds with ease. However, actually achieving that experience usually means adopting much clunkier hardware options with smart glasses for AR and larger headsets for VR, like Oculus Quest 2. The InWith soft contacts serve as an exemplar for evolving user expectations because they would provide an easier, more seamless, means to access information.

Fabric has been used to chart the smart contact lens industry (also known as bionic contact lenses) with artifacts dating as far back as 2010. One artifact from 2012 (see Figure 3.6) features the famous futurist Michio Kaku, who says that the "internet will be in our contact lens" and that we will "simply blink and be online," and that this will assist with communication, identification, and language translation. A decade of research and technical innovation has contributed to this emergent augmentation technology. Studying smart contact lens artifacts along a timeline in Fabric helps to reveal how the InWith contact lens is the first to be used in combination with Metaverse in 2022. The convergence of the smart contact lens industry and Metaverse suggests a generative enculturation process at work. It signals how companies are using Metaverse hype to encourage adoption and eventually adaptation of both concepts. It also provides technological grounding for the infrastructure that will be needed to make smart contact lenses usable.

Another major hurdle to actually creating a Metaverse is current internet infrastructure. Both authors have been researching AR as an important medium of communication for the TPC industry along with many collaborators (Duin et al., 2020; Pedersen et al., 2021). AR is a method of human–computer interaction that blends virtual representations with physical spaces, seeking to augment, mix, or shift reality (Azuma, 1997; Pedersen, 2013). After undergoing 20 years of emergence, AR is predicted to progress to become one part

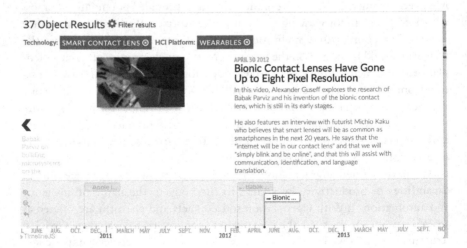

FIGURE 3.6 Smart contact lens timeline featuring a 2012 video artifact. Used with permission of Isabel Pedersen.

of a full-scale ambient platform, helping to constitute what is recognized as a Metaverse. The transformative quality of AR is that it combines virtual components with human sensory modalities (e.g., sight, sound, touch), meaning that it can enhance the first-person point of view of the user with the potential to achieve a significant sense of virtual presence.

Metaverse technologies will have to achieve interoperability for data exchange over spatial computing platforms. One technology that has been evolving for some time is AR cloud, a speculative plan to create a shared multi-user environment, a global augmented reality, and "a real time spatial map of the world" (AWE, 2018). For example, the AR cloud requires a persistent 3D digital copy of the world, which means that if a developer creates a virtual overlay for a real space, it will remain accessible to anyone across different devices and platforms. As an imagined technology future, an AR cloud could both effect positive societal change (e.g., Open AR Cloud Association) and result in greater risk through further conditions of surveillance, corporate control, or harms to humans.

Immersive technologies are already being commandeered to enable social and cultural change for the better. Lili Yan et al. (2020) argue that "immersive technology has the potential to represent the concept of the Anthropocene by engaging the user with the messages rhetorically presented." They further explain the utility of AR:

> The concept of the Anthropocene criticizes the anthropocentric human-nonhuman relationship by focusing on the interconnectedness between human and nonhuman. In this sense, the Anthropocene is a deeply cultural phenomenon... a small number of AR applications have emerged to support the understanding of the Anthropocene, the key tenets of which are usually rhetorically communicated.
>
> *(p 109)*

In short, reconceptualizing augmentation technologies through an ethical framework requires us to acknowledge and explore the ways that we are enculturated to adopt them. By doing so, we cultivate digital literacy (see Chapter 4); moreover, through strategic pedagogical use of Fabric of Digital Life artifacts, we increase understanding of AI techniques, functional applications, and AI application fields (see Chapter 6).

References

Altman, A. (2022). *InWith says it's developed world's first soft electronic contact lens*. CNET. https://www.youtube.com/watch?v=rxjcjG9XXII

Amelia Integrated Platform. (2022). https://amelia.ai/conversational-ai/

Andrejevic, M. (2020). *Automated media*. Taylor & Francis.

ATi Studios (2018, March 12). MondlyAR – learn languages in augmented reality. *Fabric of Digital Life*. https://fabricofdigitallife.com/Detail/objects/3823

AWE (Inbar, O.) (2018, January 5). *Introduction to the AR cloud* [Video]. YouTube. https://www.youtube.com/watch?v=TAlJ5t73Rr8

Ayanoglu, H., & Duarte, E. (Eds.). (2019). *Emotional design in human-robot interaction: Theory, methods and applications.* Springer.

Azuma, R. (1997). A survey of augmented reality. *Presence, 6*(4), 355–385.

Bandura, A. (2006). Toward a psychology of human agency. *Perspectives on Psychological Science, 1*(2), 164–180.

Bernstein, J. (1981, December 7). Marvin Minsky's vision of the future. *New Yorker.* https://www.newyorker.com/magazine/1981/12/14/a-i

Big tech's private passions. (2022, January 22). *Economist, 442*(9280), 18–20.

Big tech's supersized ambitions: From metaverses to quantum computing. (2022, January 22). *Economist, 442*(9280), 10.

Braidotti, R. (2021). Posthuman feminism and gender methodology. In J. Browne (Ed.), *Why gender?* (pp. 101–125). Cambridge University Press.

Building a metaverse with Chinese characteristics. (2022, February 5). *Economist, 442* (9282), 51–52.

Cairns, R., & Tham, D. (2021, August 25). More than 50 robots are working at Singapore's high-tech hospital. *CNN.* https://www.cnn.com/2021/08/25/asia/cgh-robots-healthcare-spc-intl-hnk/index.html

Campbell, K. K. (2005). Agency: Promiscuous and protean. *Communication and Critical/Cultural Studies, 2*(1), 1–19.

CanBot (2018, December 12). Xiaoyou robots made by Canbot. *Fabric of Digital Life.* https://fabricofdigitallife.com/Detail/objects/4296

Chen, B. X. (2021, January 6). The tech that will invade our lives in 2021. *New York Times.* https://www.nytimes.com/2021/01/06/technology/personaltech/tech-2021-augmented-reality-chatbots-wifi.html

Cohen, K. R. (2022). *Agent/agency.* Chicago School of Media Theory. https://lucian.uchicago.edu/blogs/mediatheory/keywords/agentagency/

Crawford, K. (2021). *Atlas of AI: Power, politics and the planetary costs of artificial intelligence.* Yale University Press.

Darling, K. (2014). Extending legal protection to social robots: The effects of anthropomorphism, empathy, and violent behavior towards robotic objects. In R. Calo, M. Froomkin, & I. Kerr (Eds.), *Robot law* (pp. 212–232). Camberley: Edward Elgar.

Duin, A. H., Armfield, D., & Pedersen, I. (2020). Human-centered content design in augmented reality. In G. Getto, N. Franklin, S. Ruszkiewicz, & J. Labriola (Eds.), *Context is everything: Teaching content strategy* (pp. 89–116). Taylor & Francis.

Duin, A. H., & Pedersen, I. (2023). *Agency, affordances, and enculturation of augmentation technologies* [Multimedia collection]. Fabric of Digital Life. https://fabricofdigitallife.com/Browse/objects/facets/collection:71

Fernandes, J. M., Sá Silva, J., Rodrigues, A., & Boavida, F. (2022). A survey of approaches to unobtrusive sensing of humans. *ACM Computing Surveys, 55*(2). https://doi.org/10.1145/3491208

Foucault, M. (1982). Afterward: The subject and power. In H. Dreyfus, & P. Rabinow (Eds.), *Michel Foucault: Beyond structuralism and hermeneutics* (pp. 208–228). University of Chicago Press.

Fove, Inc. (2015, June 10). FOVE VR headset: Tracks subtle eye movements in virtual reality. *Fabric of Digital Life.* https://fabricofdigitallife.com/Detail/objects/1103

Graham, S. S., & Herndl, C. G. (2011). Talking off-label: The role of *stasis* in transforming the discursive formation of pain science. *Rhetoric Society Quarterly, 42*(2), 145–167.

Grassia, F. S., & ElKoura, G. (2020) *Universal scene description.* CGW. https://www.cgw.com/Publications/CGW/2020/Edition-2-2020/Universal-Scene-Description.aspx

Hallenbeck, S. (2012) User agency, technical communication, and the 19th-century woman bicyclist. *Technical Communication Quarterly, 21*(4), 290–306. https://doi.org/10.1080/10572252.2012.686846

Hanson Robotics (2019, January 30). Little Sophia by Hanson Robotics. *Fabric of Digital Life.* https://fabricofdigitallife.com/Detail/objects/3541

Hayles, N. K. (2002). Flesh and metal: Reconfiguring the mindbody in virtual environments. *Configurations, 10*(2), 297–320.

Hayles, N. K. (2017). *Unthought: The power of the cognitive nonconscious.* University of Chicago Press.

Heidegger, M. (1977). *The question concerning technology and other essays.* Harper & Row.

InWith at CES 2022. (2022). *InWith to show off the ultimate metaverse wearable at CES 2022.* InWith. https://inwithcorp.com/inwith-at-ces-2022/

Iris.AI (2020, April 17). Meet AI XPRIZE Semifinalist Iris.AI. *Fabric of Digital Life.* https://fabricofdigitallife.com/Detail/objects/4883

Kass, M. (2021, September 29). *Plumbing for the metaverse with universal scene description (USD).* Nvidia Omniverse, Medium. https://medium.com/@nvidiaomniverse/plumbing-for-the-metaverse-with-universal-scene-description-usd-856a863d9b12

Kessler, M. M., & Graham, S. S. (2018). Terminal node problems: ANT 2.0 and prescription drug labels. *Technical Communication Quarterly, 27*(2), 121–136. https://doi.org/10.1080/10572252.2018.1425482

Latour, B. (1999). On recalling Ant. *The Sociological Review, 47*(1), 15–25. https://doi.org/10.1111/j.1467-954X.1999.tb03480.x

MacManus, R. (2022, January 10). Nvidia announces expansion of omniverse to consumer internet. *The New Stack.* https://thenewstack.io/nvidia-announces-expansion-of-omniverse-to-consumer-internet/

MacManus, R. (2009, July 20). The wearable internet will blow mobile phones away. *New York Times.* https://archive.nytimes.com/www.nytimes.com/external/readwriteweb/2009/07/20/20readwriteweb-the-wearable-internet-will-blow-mobile-phone-6283.html

Market capitalization of Amazon (AMZN). (2022). https://companiesmarketcap.com/amazon/marketcap/

Marr, B. (2020, September 14). The 5 biggest technology trends in 2021 everyone must get ready for now. *Forbes.* https://www.forbes.com/sites/bernardmarr/2020/09/14/the-5-biggest-technology-trends-in-2021-everyone-must-get-ready-for-now/?sh=68db4e981b82

McLuhan, M. (1964). *Understanding media: The extensions of man.* Signet.

Miettinen, R. (1999). The riddle of things: Activity theory and actor-network theory as approaches to studying innovations. *Mind, Culture, and Activity, 6*(3), 170–195.

Mosco, V. (2017). *Becoming digital: Toward a post-Internet society.* Emerald Publishing.

Neff, G., & Nagy, P. (2019). Agency in the digital age: Using symbiotic agency to explain human-technology interaction. In Z. Papacharissi (Ed.), *A networked self and human augmentics, artificial intelligence, sentience* (pp. 97–107). Taylor & Francis.

OpenAI Charter. (2018). OpenAI. https://openai.com/charter/

Oudshoorn, N., & Pinch, T. (Eds.). (2003). *How users matter: The co-construction of users and technologies.* MIT Press.

Papacharissi, Z. (Ed.). (2019). *A networked self and human augmentics, artificial intelligence, sentience.* Taylor & Francis.

Pedersen, I. (2013). *Ready to wear: A rhetoric of wearable computers and reality-shifting media.* Parlor Press.

Pedersen, I., & Duin, A. H. (2022). AI agents, humans and untangling the marketing of artificial intelligence in learning environments. *Proceedings of the 55th Hawaii International Conference on System Sciences: Human and Artificial Learning in Digital and Social Media.* https://scholarspace.manoa.hawaii.edu/server/api/core/bitstreams/522ac070-0c6a-4807-93c4-ba7d2d7f3d8a/content

Pedersen, I., Duin, A. H., Iliadis, A., & Efrat, L. (2021). *The AR cloud: Tech imaginaries, future risks, and potential affordances.* Society for Social Studies of Science Annual Meeting, October 6, 2021, Toronto, ON, Canada.

Pickering, A. (1995). *The mangle of practice: Time, agency, and science.* Chicago: University of Chicago Press.

Pixar (2022, March). *Introduction to USD.* https://graphics.pixar.com/usd/release/intro.html

Potts, L., & Jones, D. (2011). Contextualizing experiences: Tracing the relationships between people and technologies in the social web. *Journal of Business and Technical Communication, 25*(3), 338–358.

Raisamo, R., Rakkolainen, I., Majaranta, P., Salminen, K., Rantala, J., & Farooq, A. (2018). Human augmentation: Past, present and future. *International Journal of Human-Computer Studies, 131,* 131–143.

Reichardt, J. (1978). *Robots: Fact, fiction and prediction.* London: Thames & Hudson.

Russell, S., & Norvig, P. (1995). *Artificial intelligence: A modern approach.* Prentice Hall.

Schumpeter: Lords of the metaverse. (2021, December 18). *Economist, 441*(9276), 56.

Simonite, T. (2019, February 11). Trump's plan to keep America first in AI. *Wired.* https://www.wired.com/story/trumps-plan-keep-america-first-ai/

Spinuzzi, C. (2007). Who killed Rex? Tracing a message through three kinds of networks. In M. Zachry, & C. Thralls (Eds.), *Communicative practices in workplaces and the professions: Cultural perspectives on the regulation of discourse and organizations* (pp. 45–66). Routledge.

Stern, J. (2021, November, 12). Trapped in the Metaverse: Here's What 24 Hours in VR Feels Like [Video]. *The Wall Street Journal.* https://www.wsj.com/articles/metaverse-experience-facebook-microsoft-11636671113

Walton, R., Moore, K., & Jones, N. (2019). *Technical communication after the social justice turn.* Taylor & Francis.

Waverly Labs (2019, May 20). AMBASSADOR by Waverly Labs. Fabric of Digital Life. https://fabricofdigitallife.com/Detail/objects/4529

Winsor, D. (2006). Using writing to structure agency: An examination of engineers' practice. *Technical Communication Quarterly, 15*(4), 411–430.

WIPO. (2019). *WIPO technology trends 2019: Artificial intelligence.* World Intellectual Property Organization. https://www.wipo.int/edocs/pubdocs/en/wipo_pub_1055.pdf

XPRIZE (2021, June 4). *Human augmentation: Where it's at and where it's headed.* https://www.xprize.org/articles/human-augmentation-where-it-s-at-and-where-it-s-headed

Yan, L., Colleni, M., & Litts, B. K. (2020). *Exploring the rhetorical affordances of augmented reality in the context of the Anthropocene.* 2020 6th International Conference of the Immersive Learning Research Network (iLRN), 109–116. http://dx.doi.org/10.23919/iLRN47897.2020.9155126

SECTION TWO
Build Literacies

Section Two provides detailed definition of digital and AI literacy amid the emergence of augmentation technologies, with focus on examining, exploring, and participating in the development of digital and AI literacy skills needed to address algorithmic mining and bias, racial discrimination, digital divides, unethical AI practices, misinformation, and other socio-ethical harms to humans.

DOI: 10.4324/9781003288008-5

4

COMPETENCIES, DESIGN CONSIDERATIONS, AND NEW ROLES FOR WORK WITH AUGMENTATION TECHNOLOGIES AND AI

Overview

Chapter 4 expands and clarifies the definition of digital and AI literacy as a means to build literacy for current and future work with augmentation technologies. The exponential increase in AI techniques, functional applications, and use across application fields demands critical attention to digital literacy, data literacy, AI literacy, AI explainability, and Trustworthy AI. This chapter uses Long and Magerko's (2020) definition of AI literacy and conceptual framework for determining AI competencies and design considerations as a means for technical and professional communication (TPC) scholars, instructors, students, and practitioners to examine and develop digital and AI literacies. It focuses on Explainable AI (XAI) to help humans better understand how AI works and makes specific decisions. It then discusses the European Union's Assessment List for Trustworthy AI (ALTAI, 2020) strategic questions for use as an initial approach to evaluating Trustworthy AI for the purpose of minimizing risks while maximizing the benefits of AI for human users. The chapter concludes with discussion of how TPC roles are evolving as a result of augmentation technologies and AI.

Key Questions

- What competencies enable the ability to critically evaluate AI technologies as well as communicate and collaborate effectively with them?
- How might TPCs help humans to better understand how AI works and makes decisions?

DOI: 10.4324/9781003288008-6

- How might TPCs determine trustworthiness of AI for the purpose of minimizing risks while maximizing the benefits of AI for human users?
- How will TPC roles change as a result of AI techniques, functional applications, and use across multiple fields?

Chapter 4 Links

Throughout the chapter, we refer to articles, videos, and reports which can be found in a related chapter collection at Fabric of Digital Life called *Competencies, design considerations, and new roles for work with augmentation technologies and AI* (Duin & Pedersen, 2023). You can find a link to this collection at https://fabricofdigitallife.com.

Introduction

In the summer of 1961, Simulmatics Corporation scientists, "Cold War America's Cambridge Analytica," hunkered down to plan new projects for their invention. As historian Jill Lepore (2020a) chronicles, this invention was "a computer program designed to predict and manipulate human behavior, all sorts of human behavior, from buying a dishwasher to countering an insurgency to casting a vote. They called it the People Machine" (p. 2). The underlying proposition guiding these social scientists was the belief that if they collected enough data about people and fed it into this machine, then "everything, one day, might be predictable, and everyone, every human mind, simulated, each act anticipated, automatically, and even driven and directed, by targeted messages as unerring as missiles" (pp. 3–4). Staffed by eminent scientists, Simulmatics gathered opinion-poll data to create its models of the US electorate. Lepore (2020b), writing in *Nature*, details the eruption of a story in *Harper's* magazine about the People Machine and its mysterious "What-If Men" influence on the 1960 US presidential election:

> The ensuing debate raised questions that are still asked today—urgently. Can computers rig elections? What does election prediction mean for democracy? What does automation mean for humanity? What happens to privacy in an age of data? There were no answers then, as now.
>
> *(p. 349)*

In 1969, the Simulmatics Corporation began to unravel amid student protests regarding its use for work in Vietnam. Its most direct descendant is the MIT Media Lab, where exploration of similar technologies continues.

Fast forward to the present. As Lepore writes, "Facebook, Palantir, Cambridge Analytica, Amazon, the Internet Research Agency, Google—they were all incubated there" (p. 4). Decades of data, old and new, predict and anticipate behavior. "Strategically hidden inside our computers are files that track our movements on the web... long strings of alphanumeric codes" (Beck, 2015, p. 125). The BlueKai Registry program that technical communication researcher Estee Beck used to examine her digital identity representation (http://www.bluekai.com/registry/) no longer exists; this link now accesses Oracle Advertising (2022): "In an increasingly complex and connected world, you need to know who your best audiences are and be able to reach them in the right environment" (https://www.oracle.com/cx/advertising/). The power, the predictability, has transitioned from individuals to large corporations. As contributors to Estee Beck and Les Hutchinson Campos's (2020) collection attest, surveillance and datafication similarly engulf the teaching of writing.

Defining Digital, Data, and AI Literacy

From 1980 to 1999, six literacy terms appeared synonymously across scholarship: "information literacy," "computer literacy," "library literacy," "media literacy," "network literacy", and "digital literacy" (Bawden, 2001, p. 219). David Bawden's tracing of the occurrence of these terms found "digital literacy" appeared the least, only first identified in 1997 as a way to describe the shifting nature of reading and writing given mediation by the internet. Bawden's (2008) later work chronicles the origins and concepts of digital literacy, beginning with Paul Gilster's (1997) explanation of it "as an ability to understand and to use information from a variety of digital sources... the ability to read, write and otherwise deal with information using the technologies and formats of the time—an essential life skill." Gilster states this explicitly: "digital literacy is about mastering ideas, not keystrokes" (as discussed by Bawden, 2008, p. 18).

Digital Literacy

In 2008, Colin Lankshear and Michele Knobel defined two main kinds of digital literacy: *conceptual definitions* and *standardized sets of operations* or skills. For conceptual definitions, they noted Richard Lanham's (1995) claim that being digitally literate involves "being skilled at deciphering complex images and sounds as well as the syntactical subtleties of words" (p. 200). Lanham noted that digitally literate people are:

> quick on [their] feet in moving from one kind of medium to another... know what kinds of expression fit what kinds of knowledge and become skilled at presenting [their] information in the medium that [their] audience will find easiest to understand.

In this same time period, Paul Gilster (1997) also defined digital literacy as including the ability to understand and use information in multiple formats, identifying four digital literacy competencies: knowledge assembly, evaluating information content, searching the internet, and navigating hypertext. For standardized sets of operations, Lankshear and Knobel noted the Global Digital Literacy Council's (GDLC, 2022) set of Digital Literacy Standards and subsequent Computing Fundamentals' test items as well as the US Educational Testing Service (ETS) definition of digital literacy as "the ability to use digital technology, communication tools and/or networks appropriately to solve information problems in order to function in an information society" (https://ets.org). Such ability comprises "the ability to use technology as a tool to research, organize, evaluate, and communicate information, and the possession of a fundamental understanding of the ethical/legal issues surrounding the access and use of information" (Educational Testing Service, 2003, p. 11).

Subsequent definitions of digital literacy parallel this framework of skills paired with responsible use of technology. According to the Museum and Library Services Act of 2010 (Pub.L. 111–340, 22 Dec. 2010), digital literacy means "the skills associated with using technology to enable users to find, evaluate, organize, create, and communicate information; and developing digital citizenship and the responsible use of technology." The American Library Association (ALA, 2022) defines digital literacy as "the ability to use information and communication technologies to find, evaluate, create, and communicate information, requiring both cognitive and technical skills." An Issue Brief on Digital Literacy from the American Institutes of Research (Vanek, 2020) both reiterates and expands this definition, stating that:

> the crux of what is meant by digital literacy is the recognition of these skills' relevance in specific contexts and one's ability to creatively apply them (International Society for Technology in Education, 2016; Jacobs & Castek, 2018; Vanek, 2017). Also important to note, digital literacy is often referred to as one monolithic construct, but it is really one that encompasses several groups of competencies.

In *Writing Futures* (2021), we emphasized the evolving literacy as human and machine work together:

> Digital literacy for writing futures means no longer viewing human and machine as separate agents along with the ability to envision and write within mirror-worlds of virtual fragments stitched together. It requires reaching beyond current copresent relationships with agents through devices and fostering the ability to
>
> • Expand digital literacy to include visual literacy through work with augmented and virtual reality;

- Understand technological embodiment, agency developed out of human and nonhuman collaboration—for example, how one's writing capacities are extended through and with device collaboration;
- Identify collaboration-enabling features—for example, articulating the ways that human–nonhuman collaboration shapes communicative actions and interactive behaviors;
- View human–machine collaboration as a dialogic conversation, assuming that a collaborative, integrated relationship will develop between human and nonhuman agent, leading to greater knowledge by both human and agent; and
- Project how one's body adapts to networks, assemblages, or even as a host for future human and nonhuman collaboration in ambient interactive relationships (p. 40).

Numerous worldwide initiatives document digital literacy competencies for personal, professional, and civic engagement. Western Sydney University's Library Study Smart (2022) resources define digital literacy as "having the skills you need to live, learn, and work in a society where communication and access to information is increasingly through digital technologies." This set of resources includes seven categories and associated competencies: digital literacy definition, ICT (Information and Communication Technology) proficiency, information literacy, digital scholarship, communications and collaborations, digital learning, and digital identity. Likewise, based on an extensive review of articles, reports, frameworks, specifications, and standards as well as interviews, the UK's Joint Information Systems Committee (Jisc, 2020) Digital Capability Framework leadership identified key issues in framing how to deepen digital know-how, defining digital literacies as "the capabilities which fit someone for living, learning and working in a digital society." In this framework digital literacy competencies include ICT proficiency; data and media literacies; digital creation, problem solving, and innovation; digital communication, collaboration, and participation; digital learning and development; and digital identity and wellbeing. And the worldwide Global Digital Literacy Council (https://www.gdlcouncil.org/) establishes standards for validation of digital competencies. However, to our knowledge, no initiative focuses on examining and/or curating artifacts on emerging technologies to actively engage in building digital literacy.

To address this gap, since 2019 and together with Jason Tham, we have co-directed the international *Building Digital Literacy* (BDL) project for the purpose of building digital literacy through examining and/or curating collections of emerging technologies at the Fabric of Digital Life (https://fabricof-digitallife.com/). Over the course of the project, instructors have met every two weeks across multiple institutions and have collaborated on joint problem solving to build collective understanding of digital literacy; design and deploy sets of instructional materials for student exploration; curate collections in this repository; and conduct studies on the impact of such work in building digital literacy. As these instructor-scholars reviewed documents, they participated in ongoing discussion of digital literacy. As one example, BDL scholars used Peter Stordy's (2015) taxonomy of literacies and definition of digital literacy to develop new directions for instructional units and student examination and curation of Fabric artifacts. Stordy defines digital literacy as "the abilities a person or social group draws upon when interacting with digital technologies to derive or produce meaning, and the social, learning and work-related practices that these abilities are applied to" (p. 472).

We have studied the collaborative engagement in place to foster and support instructional development, pedagogical deployments, and associated research in support of building student digital literacy (Duin et al., 2021). From our primary discipline, TPC, we note the following:

> Digital literacy scholarship has revolved mainly around the use of computers for composing and producing meaning (Breuch, 2002; Hovde & Renguette, 2017; Selber, 2004). Technical communication scholars have focused on technological literacy (tool knowledge), and most recently, code literacy (content management rules/knowledge). Hovde and Renguette (2017), drawing on the work of technical communication scholars who have addressed technological literacy (Breuch, 2002; Brumberger et al., 2013; Cargile Cook, 2002; Northcut & Brumberger, 2010; Selber, 2004; Turnley, 2007), consolidate this scholarship into functional, conceptual, evaluative, and critical levels of technological literacy. Instructional models in technical communication exist for the purpose of cultivating technological and most recently, code literacy (Duin & Tham, 2018). However, no innovative model exists for building digital literacy, i.e., literacy that includes 'making ethically informed choices and decisions about digital behaviour… digital safety, digital rights, digital property, digital identity and digital privacy' (Traxler, 2018, p. 4).
>
> *(Duin et al., 2021, p. 94)*

Open to all, this BDL project's joint problem solving continues to focus on the development of instructional materials and guided prompts for use across deployments as a means to foster student development of digital literacy (see Chapter 6).

Data Literacy

The amount of data surging amid human and machine collaboration and within augmentation technologies demands critical attention to data and AI literacy. The Gartner Corporation defines data literacy as "the ability to read, write and communicate data in context, including an understanding of data sources and constructs, analytical methods and techniques applied, and the ability to describe the use case, application and resulting value" (Panetta, 2021). To assess data literacy in an organization, they suggest these questions to quantify data literacy:

- How many people in your business can interpret straightforward statistical operations such as correlations or judge averages?
- How many managers are able to construct a business case based on concrete, accurate and relevant numbers?
- How many managers can explain the output of their systems or processes?
- How many data scientists can explain the output of their machine learning algorithms?
- How many of your customers can truly appreciate and internalize the essence of the data you share with them? (Panetta, 2021)

MIT professor Catherine D'Ignazio and research scientist Rahul Bhargava (2015) describe data literacy as the ability to do the following:

- Read data, which means understanding what data is and the aspects of the world it represents;
- Work with data, including creating, acquiring, cleaning, and managing it;
- Analyze data, which involves filtering, sorting, aggregating, comparing, and performing other analytic operations on it; and
- Argue with data, which means using data to support a larger narrative that is intended to communicate some message or story to a particular audience.

Somewhat in contrast, Luci Pangrazio and Neil Selwyn (2019), arts and education faculty members in Australia, define data literacy as the ability to understand and protect one's personal data, emphasizing that data literacy is now more crucial than digital literacy: "traditional concerns over supporting the development of 'digital literacy' are now being usurped by concerns over citizens' 'data literacies'" (p. 419). They propose a framework of "Personal Data Literacies" distinguished by five domains: (1) Data Identification, (2) Data Understandings, (3) Data Reflexivity, (4) Data Uses, and (5) Data Tactics (p. 419). They state that:

> developing understandings that are not only more technically accurate but also more attuned to the complex and evolving socio-technical processes and systems underpinning contemporary digital society is fundamental to practical efforts to support personal data education, as well as research into everyday uses of data-based technologies.
>
> *(p. 420)*

We strongly concur. To develop such understanding requires critical attention to AI literacy.

Global literacy projects are paying attention as well. Partnering companies have launched the global Data Literacy Project (https://thedataliteracyproject.org/) to "ignite discussion" and develop tools to build a digitally literate society. Moreover, the European Commission's DigComp 2.1 project presents a Digital Competence Framework for Citizens that integrates basic digital literacy competencies (information and data literacy, communication and collaboration, digital content creation, safety, and problem solving) and basic data literacy competencies (data collection, data management, data application, and data evaluation) as they interact seamlessly and concurrently (Carretero Gomez et al., 2017). Each of these initiatives emphasizes AI literacy.

AI Literacy

MIT Ph.D. candidate in Medical Engineering and Medical Physics and science communicator, Jordan Harrod (2020) asks, "What does it mean to be literate in AI?" and she responds that "it's the knowledge that a person accumulates in order to be able to confidently understand and interact with AI based systems." Regarding the goal for AI literacy, Harrod emphasizes that:

> we don't necessarily focus on understanding the specifics of the hardware in AI literacy; we're not necessarily concerned with the specifics of the code; instead, we're interested in understanding how these algorithms affect our daily lives and how we could use them for our own benefits.

In terms of how algorithms affect our daily lives, in *Writing Futures* (2021), we focused on how human writers will collaborate with AI autonomous agents and the importance of cooperative scenarios among humans and machines. We highlighted the importance of natural language generation (NLG), a subfield of AI concerned with computer systems, e.g., machine learning (ML) systems that produce understandable texts. Fundamental to understanding AI literacy is the understanding of text-to-text and data-to-text generation. In text-to-text generation, software uses publicly available digital information, including human-written texts, and applies algorithms to create new texts. Examples range from automatic spelling and grammar correction to peer reviews for scientific papers. In data-to-text generation, software converts data into texts without using previous texts written by human writers. Examples here range from weather and financial reports to comments intended to persuade and motivate behavior. We also shared multiple links to automated writing programs, discussed implications including robot or algorithmic journalism, and commended initiatives, including the Finnish government's Elements of AI course in collaboration with the University of Helsinki and Reaktor (University of

Helsinki, 2022). Its goal is to demystify AI and encourage citizens to learn what AI is, what can and cannot be done with it, and how to create methods for continuing to understand AI as it evolves. These programs provide for greater understanding of AI; however, they do not specify competencies for understanding and using AI.

In Chapter 2, we highlighted the importance of international multi-stakeholder organizations (e.g., WHO, WIPO, UN) that orient augmentation technologies toward a collective global worldview for inclusive ends, and in Chapter 3 we highlighted the World Intellectual Property Organization's (WIPO) Technology Trends 2019 report's use of the three categories: AI techniques, AI functional applications, and AI application fields. In this chapter we expand understanding of AI literacy by discussing key themes from multiple fields that further define AI literacy, with focus on component competencies. To do so, we highlight the scholarship of Duri Long and Brian Magerko (2020) at the Georgia Institute of Technology, who conducted an exploratory review of 150 documents primarily from 2000 to 2019 to distill key ideas from multiple fields that inform understanding of how learners make sense of AI. Resources included proceedings from the Association for the Advancement of AI's (AAAI) AI Education Colloquium (AI Ed); proceedings of post-2016 conferences including AAAI, AI Ed, CHI (Computer Human Interaction), and IDC; searches using Google Scholar and the ACM digital library; 14 public syllabi from accredited public USA universities for classes related to AI, ML, cognitive science, and robotics; contents of popular AI textbooks; peer-reviewed literature on perceptions of AI; and AI education literature.

We agree with their resulting definition of AI literacy: "a set of competencies that enables individuals to critically evaluate AI technologies; communicate and collaborate effectively with AI; and use AI as a tool online, at home, and in the workplace" (p. 2). Note that one could add "augmentation technologies" made possible by AI throughout this definition. We continue with an overview of Long and Magerko's research questions along with their detailed set of 17 competencies and 15 design considerations (see Table 4.1).

"What Is AI?"

Long and Magerko's first research question, "What is AI?" includes four competencies: Recognizing AI, Understanding Intelligence, Interdisciplinarity, and General vs. Narrow (p. 4). Determining what constitutes AI involves recognizing when AI is in use along with the ability to analyze and discuss features that make an entity intelligent. For a pedagogical example, Abby Koenig (2020) provides opportunities for TPC students to engage with algorithmic platforms (Amazon, Facebook, Instagram, Twitter, Match.com, etc.), using student journals to explore and reflect on algorithmic engagements. Results from her study indicate initial student understanding of AI to be "surface level." Students' later

TABLE 4.1 AI research questions, competencies, and associated design considerations (Long & Magerko, 2020)

Research questions	Competencies	Design considerations
What is AI?	Recognizing AI	
	Understanding intelligence	
	Interdisciplinarity	
	General vs. narrow	
What can AI do?	AI strengths and weaknesses	
	Imagine future AI	
How does AI work?	Representations	Explainability
	Decision-making	Embodied interactions
	Machine learning (ML) steps	Contextualizing data
	Human role in AI	
	Data literacy	
	Learning from data	
	Critically interpreting data	
	Action and reaction	
	Sensors	
How should AI be used?	Ethics	
How do people perceive AI?	Programmability	Promote transparency
		Unveil gradually
		Opportunities to program
		Milestones
		Critical thinking
		Identity, values, & backgrounds
		Support for parents
		Social interaction
		Leverage learners' interests
		Acknowledging preconceptions
		New perspectives
		Low barrier to entry

reflections on algorithmic engagement indicate more critical and rhetorical responses to algorithms. Koenig concludes:

> When offered the opportunity to explore the role the platforms play in their lives, they became more critical and rhetorically aware of just how influential the platforms were. When students reflected in their journals, there were moves towards critical or rhetorical awareness that I did not find in opening paragraphs of the journals. While these moves were not monumental, the data supported that these moves were notable and repeated throughout each of the four journal prompts. I argue that being asked to write about their algorithmic engagements opened a door

towards a different type of awareness, and the potential for them to become more critical and rhetorical actors within their algorithmic engagements was possible.

This example assignment also assists in analyzing and discussing features that make an algorithmic platform intelligent as well as the many types of systems (cognitive, sensory, physical, emotional) that use AI.

Regarding what constitutes AI, Long and Magerko also note the need to distinguish between general and narrow AI, with narrow AI being that which is intelligent in a particular domain, and general being AI that which "rivals human intelligence across multiple domains" (p. 4). Narrow AI systems are good at a single or limited range of tasks, but once outside of that range, they fail. Ben Dickson (2020) offers this example:

> a bot developed by the Google-owned AI research lab DeepMind can play the popular real-time strategy game StarCraft 2 at championship level. But the same AI will not be able to play another RTS game such as Warcraft or Command & Conquer.

General AI or artificial general intelligence (AGI):

> would be a machine capable of understanding the world as well as any human, and with the same capacity to learn how to carry out a huge range of tasks.... [It] would be able to combine human-like, flexible thinking and reasoning with computational advantages, such as near-instant recall and split-second number crunching.
>
> *(Heath, 2018)*

AGI is progressing through natural language processing (NLP), an AI technology that comprehends human language, reducing the human need to interact with a screen; through rapid development of the Metaverse as companies develop immersive technologies for work in this virtual world; through development of low-code to no-code AI for building applications and conversational AI chatbots or AI-enabled virtual assistants as customer support agents; and via technology such as Quantum AI to analyze huge volumes of data in minimal time (Kanade, 2022).

"What Can AI Do?"

Long and Magerko's second research question, "What can AI do?" includes two competencies: AI's Strengths & Weaknesses, and Imagine Future AI. As a means to determine what AI can do and to imagine its futures, throughout this book we encourage readers to use the Fabric of Digital Life cultural archive to examine technologies that exhibit AI strengths and weaknesses and to consider the meta-data behind each artifact in order to chronicle the past, present, and emerging future of each artifact. As AI fuels augmentation technologies, again consider the four enhancements—cognitive, emotional, sensory, and physical—to determine what AI can do and how it's being imagined for the future (see Chapter 2).

As an example of what generative AI can do, OpenAI's ChatGPT was released at the time of publication of this book. Co-writing content with AI eclipses the older notion of AI assistantship. As AI technologies are designed to assume the agentive roles of editorship and authorship, we need to re-envision the way we conceptualize writing with AI agents. With OpenAI's suite of Large Language Models (LLMs), human-machine synergy becomes operational.

"How Does AI Work?"

Their third research question, "How does AI work?," includes nine competencies and three design considerations. These competencies include elements of digital and data literacy as a means to recognize the human role in AI and to understand basic data literacy concepts and the need to critically interpret the data. Long and Magerko's prescription to have learners investigate who created the dataset, how the data was collected, and limitations of the dataset can be implemented across a variety of pedagogical contexts and made mandatory across numerous professional contexts. Furthermore, their note of the need to recognize sensors aligns with current TPC research based on postconnectivist technical communication for a datafied world (Verhulsdonck & Tham, 2022) in which they investigate human agency amid automated technological infrastructures in cities in the US, Europe, and Asia.

"How Should AI Be Used?"

Their fourth research question, "How should AI be used?," includes one core competency, Ethics. They emphasize the importance of identifying and describing perspectives on AI ethical issues including "privacy, employment, misinformation, the singularity, ethical decision-making, diversity, bias, transparency, accountability" (p. 7). Chapter 5 focuses on these socio-ethical consequences of augmentation technologies.

Both questions—"How does AI work?" and "How should AI be used?"—focus on AI representations, AI data, and critically interpreting the data. In terms of the competency Human Role in AI, Jeffrey Hancock et al. (2020), writing as part of the Computer-Mediated Communication (CMC) community, define Artificial Intelligence-Mediated Communication (AI-MC) as follows:

> Artificial Intelligence–Mediated Communication (AI-MC) [is] interpersonal communication in which an intelligent agent operates on behalf of

a communicator by modifying, augmenting, or generating messages to accomplish communication goals…. AI-MC raises new questions about how technology may shape human communication and requires re-evaluation—and potentially expansion—of many of Computer-Mediated Communication's (CMC) key theories, frameworks, and findings. A research agenda around AI-MC should consider the design of these technologies and the psychological, linguistic, relational, policy and ethical implications of introducing AI into human–human communication.

(p. 89)

Hancock et al. include multiple dimensions of AI-MC:

- Magnitude, i.e., the changes made by the AI agent, ranging from a word change to full content generation;
- Media type, e.g., text, audio, and video editing tools and filters to make one look more attractive, trustworthy, or similar to the receiver;
- Autonomy granted to the AI to operate on behalf of the communicator; and
- Role orientation, or whether the AI is operating on behalf of the sender or the receiver [or both] (pp. 91–92).

Hancock et al. do not consider AI-MC as related to AI-human interaction; rather, their focus is on algorithms that mediate human-AI communication.

"How Do People Perceive AI?"

Long and Magerko's final research question, "How do people perceive AI?," focuses nearly entirely on design considerations. These design considerations align closely with TPC objectives of effective information design to promote transparency and unveil information gradually; coding literacy through providing opportunities to program; attention to audience identity, values, backgrounds, and interests; designing to promote usability; and incorporating social interaction.

Consider Alan Winfield, Professor of Robot Ethics at UWE Bristol, who regularly asks this question to students, scholars, and practitioners, "How much AI is in your lives?" He then points them to the crowdsourced AI incident database (https://incidentdatabase.ai/) for concrete examples of AI failures. In describing this project, the Partnership in AI (PAI) staff (2022) write:

This project sprang directly from the needs of the PAI research community, which found there to be no prior resource offering a comprehensive overview of AI safety and fairness failures. Now, with the crowdsourced AIID, an AI researcher can type 'policing' into the database's search field and retrieve dozens of citable incident reports on the topic. Similarly, corporate product managers hoping to anticipate and mitigate risks could search the AIID for 'translate' before launching a new translation service. They then might learn about Incident 72, when a social media user was reportedly arrested after his 'good morning' message was automatically translated as 'attack them.'

Another incident (#59) shows that Google Translate systematically changes the gender of translations when they do not fit with stereotypes. Exploration of this database provides multiple examples of ML algorithms picking up "deeply ingrained race and gender prejudices concealed within the patterns of language use" (Devlin, 2017).

Explainable Artificial Intelligence (XAI)

Of note is Long and Magerko's emphasis on the design consideration explainability. Given the profound implications of explainability for technical and professional communicators, it is critical for us to add more detail regarding Explainable AI (XAI). XAI has emerged to help humans better understand how AI works and makes particular decisions. In essence, XAI provides explanations about AI's inner logic. As we increasingly collaborate with nonhuman agents, and as augmentation technologies continue to proliferate, understanding the AI on which these are built becomes more crucial. According to IBM (2022):

> Explainable artificial intelligence (XAI) is a set of processes and methods that allows human users to comprehend and trust the results and output created by machine learning algorithms. Explainable AI is used to describe an AI model, its expected impact and potential biases. It helps characterize model accuracy, fairness, transparency and outcomes in AI-powered decision making. Explainable AI is crucial for an organization in building trust and confidence when putting AI models into production. AI explainability also helps an organization adopt a responsible approach to AI development.

Matt Turek (2018) from the US Defense Advanced Research Projects Agency (DARPA) provides two aims for the Department of Defense's (DoD) Explainable AI program:

- Produce more explainable models, while maintaining a high level of learning performance (prediction accuracy); and
- Enable human users to understand, appropriately trust, and effectively manage the emerging generation of artificially intelligent partners.

In terms of scholarship, Alejandro Barredo Arrieta and colleagues (2020) analyzed 400 recent studies on XAI and related concepts to articulate this definition for XAI: "Given an audience, an **explainable** Artificial Intelligence is one that produces details or reasons to make its functioning clear or easy to understand" (p. 85). They identified the different purposes of XAI as sought by different audiences, noting two prevailing goals: the need for understanding and regulatory compliance. For each target audience, Arrieta et al. indicate why that audience needs Explainable AI:

- Domain experts such as medical doctors and insurance agents need XAI as a means to trust the model and gain scientific knowledge;
- Regulatory entities and agencies need to certify model compliance with legislative and legal requirements;
- Managers and executive board members also need to assess regulatory compliance as well as understand corporate AI applications;
- Data scientists, developers, and product owners need to ensure and improve product efficiency and functionalities; and
- Users affected by the AI system need to understand it as a means to verify fair decisions.

One might ask, should a medical doctor trust the inference of an AI model when diagnosing a disease? Consider medical use of the electronic health record giant Epic Systems Corporation (EPIC) for providing patient health care. More than 250 million US patient records are held within this system, and medical doctors increasingly rely on its predictive AI algorithms when prescribing care as well as a way to gain scientific knowledge (Adams, 2021). Its Cosmos program mines data from millions of patient medical records to improve research into treatments; however, EPIC's ML algorithms are not disclosed, nor its model development data explained (Raden, 2021). People should have the right to "Human Review of [an] Automated Decision," which is "the idea that where AI systems are implemented, people who are subject to their decisions should be able to request and receive human review of those decisions" (Fjeld et al., 2020, p. 53). In contrast, medical researchers at the University of Minnesota have partnered with EPIC to develop and validate an AI algorithm that can evaluate chest X-rays to diagnose possible cases of COVID-19 (University of Minnesota, 2022).

One might also ask, should a medical doctor trust AI race recognition from X-rays? To investigate "the underlying mechanism by which AI models can recognise race," Judy Wawira Gichoya and colleagues recently (2022) evaluated deep learning race detection models from X-rays followed by assessing the detection of race "in isolation using regression models, and by re-evaluating the deep learning models by testing them on [new] datasets." While their finding that standard AI deep learning models can be trained to predict race from

X-rays is not surprising, they also found that AI can accurately predict race "even from corrupted, cropped, and noised medical images, often when clinical experts cannot," thus creating "an enormous risk for all model deployments in medical imaging" (p. e406). Hiawatha Bray (2022), writing for the *Boston Globe*, summarizes the findings:

> the study found that an artificial intelligence program trained to read X-rays and CT scans could predict a person's race with 90 percent accuracy. But the scientists who conducted the study say they have no idea how the computer figures it out.

The risk is that doctors' decisions will be premised on biased results:

> the research raises the unsettling prospect that AI-based diagnostic systems could unintentionally generate racially biased results. For example, an AI (with access to X-rays) could automatically recommend a particular course of treatment for all Black patients, whether or not it's best for a specific person. Meanwhile, the patient's human physician wouldn't know that the AI based its diagnosis on racial data.
>
> *(Bray, 2022)*

These examples illustrate the need for regulatory agencies and board members to use XAI to understand and certify regulatory and legislative compliance; for scientists and developers to ensure and improve efficiencies and accuracies; and for end-users, e.g., health patients, to better understand their situation to verify the decisions and prescriptions made. XAI clearly is gaining traction; however, Ferreira and Monteiro (2021) emphasize the need to discuss XAI early on in the design process with both developers and end-users. Chapter 5 includes sections on fairness and non-discrimination, transparency and explainability, and accountability, which describe international principles on these topics; and Chapter 8 focuses on strategic and tactical approaches to regulation and compliance.

Designing XAI into emerging interfaces is pivotal for testing how well products communicate with humans. As an example, in *Writing Futures*, we included Pepper (see Figure 4.1), a robot from SoftBank Robotics, whose claim is to be "much more than a robot, he's a companion able to communicate with you through the most intuitive interface we know: voice, touch and emotions" (2021, p. 99). Peter Varhol (2021) from Technology Strategy Research describes how Pepper talks through its instructions as it executes them, effectively a form of XAI that helps humans understand why it is doing specific activities; it will also identify questions and ambiguities and ask for clarification. Varhol writes:

> Imagine how such a program feature can assist testers. Using test data, the tester can obtain a result, then ask the application how it obtained that

FIGURE 4.1 Pepper from SoftBank Robotics Europe. Creative Commons Attribution-Share Alike 4.0 International license. https://creativecommons.org/licenses/by-sa/4.0/

result, working through the process of manipulating the input data so that the tester can document why the result is valid.

He also emphasizes the need for XAI to serve multiple audiences: "For developers, it can help validate the technical approach and algorithms used. For testers, it helps confirm correctness and quality. For users, it is a way of establishing trust in the application." Ultimately, XAI is about establishing trust.

A resource for TPC use is the Ericsson Company's focus on establishing human trust in AI. Their article "Explainable AI—How Humans Can Trust AI" provides a solid overview of XAI, with trust being enhanced by the following seven properties, each of which represents a role for the technical and professional communicator:

- Trustworthiness, to attain the trust of humans on the AI model by explaining the characteristic and rationale of the AI output

- Transferability, where the explanation of an AI model allows better understanding of it so that it can be transferred to another problem or domain/application properly
- Informativeness, relating to informing a user regarding how an AI model works in order to avoid misconception
- Confidence, which is achieved through having a model that is robust, stable, and explainable to support human confidence in deploying an AI model
- Privacy awareness, ensuring that the AI and XAI methods do not expose private data
- Actionability, with XAI providing indications regarding how a user could change an action to yield a different outcome in addition to providing the rationale for an outcome
- Tailored (user-focused) explanations, allowing humans… to understand the behavior and predictions made by AI based systems… based on their roles, goals, and preferences (Inam et al., 2022).

The classic role of TPCs is to understand the rhetorical context, describe how something works, and make it usable. With algorithms, this role is now one of explaining how the algorithm works, changes, and adapts over time. TPCs can help users understand the overall rhetorical context as well as characteristics and rationale of AI output, and identify how its use might be transferred to another context, problem, or user need. As with all usability, it is critical for TPCs to help users avoid misconception, build confidence, and ensure privacy. Moreover, TPCs can help users to take appropriate actions to yield different AI outcomes, and they can assist in identifying and explaining how user roles, goals, and preferences influence predictions made by AI systems.

The importance of TPCs in the AI and healthcare space is critical as workers are positioned to concentrate at the top percentages of a task, with algorithmic assistance providing a first pass at a diagnosis (He et al., 2019). In the AI and healthcare space, explainability, trust, and credibility carry the consequences of being right or wrong (as in life or death). TPCs can guide an organization in developing explainability, especially at the point of introducing new capabilities in how something will work. This requires initial building of belief and credibility about where and how the AI works well along with ongoing maintenance of how it is continuing to perform. Routine work such as initial readings of X-rays saves time. The messages constructed from such readings need to be crafted for the appropriate audience and context to establish credibility. Explanation and explainability are crucial. The higher the potential risk, the greater the need for explainability.

In short, we contend that explainability and XAI should become a priority for TPCs. For XAI, we suggest reading and sharing more general overviews such as Orhan Yalcin's (2020) article on the need to seek explainability and interpretability in AI systems. He notes that while AI may well enhance lives, the increased use of AI systems trained with so-called deep learning leads to less and less explainability of ML algorithms. For example, Alberto Romero (2021) writes that Open AI's NLP model, GPT-4, will have as many parameters as the brain has synapses. Individually or as a group, you can access and use the Moral Machine platform (https://www.moralmachine.net/) to judge or browse scenarios regarding decisions made by machine intelligence such as self-driving cars. And Turek's (2018) overview of XAI shows how the military is using research on the psychology of explanation to develop new AI systems "where machines understand the context and environment in which they operate, and over time build underlying explanatory models that allow them to characterize real world phenomena."

In short, decision makers—doctors, drivers, judges, bankers, teachers—require explanations from the AI systems being used. "As AI becomes pervasive in high-stakes decision-making tasks, such as in supporting medical, military, legal, and financial judgments, making AI explainable to its users is crucial to identify potential errors and establish trust" (Jin et al., 2021, p. 1). TPCs have long known how users vary based on diverse roles, tasks, and goals. The highly technical developer utilizes explanations to debug, gain insights on the model, and improve it. The doctor needs explanations while using a diagnostic support system, the human resource specialist needs explanations to support hiring decisions, a house seller may leverage the explanation to boost property value, and the realtor may need to understand why a bid was lost.

Weina Jin and colleagues at Simon Fraser University also stress the need for end-user-centered-XAI (EUCA). EUCA includes 12 end-user-friendly explanatory forms grouped into four categories of explanation:

- Rules, e.g., a decision tree or decision rule;
- Examples, such as a similar example, typical example, or counterfactual example;
- Features, such as feature attribution, feature shape, or feature interaction; and
- Supplementary information, including input, output, datasets, and information on performance (Jin et al., 2020, p. 2).

Jin et al. write that:

> the main contribution of EUCA is that it provides a practical prototyping framework for AI practitioners (UX designers, developers, etc.) to build end-user-centered XAI prototypes. The prototyping workflow and tangible design examples/templates support user-centered prototyping and co-design process, and enable end-users to communicate their context-specific explainability needs to practitioners.
>
> *(p. 3)*

TPCs looking to explain AI to end-users will benefit from this explainability design framework.

Researchers Mireia Ribera and Agata Lapedriza (2019) also provide a review of XAI with focus on user-centered Explainable AI. They suggest classifying users into three groups—developers or AI researchers, domain experts, and lay users—and creating different explanations based on the target group. While this suggestion may appear obvious to TPCs, their discussion of explainability highlights the fact that explanation itself is ill-defined. Explainability relates to "transparency, interpretability, trust, fairness and accountability, among others" (p. 2). Of note for TPCs is their work on how the concept of the user manual might be reimagined as an explanation. If end-users are going to be involved in the training process of machines, they may need some guidance/explanation about how to do this effectively. Although much of this explanation will be built into the interaction with the AI product/system, TPCs should play a key role in this part of the design as well.

TPCs can also explore tools such as Shapash (https://shapash.readthedocs.io) that may be helpful for data scientists and that also could be utilized by TPCs contributing to the design of usable AI. Shapash offers visual summaries of algorithms using a python library, a collection of code, that "aims to make machine learning interpretable and understandable to everyone." For end-users, Shapash's web application can make it clearer how a model came to a particular decision and can help data scientists explain data results for those who don't work with data.

In the end, while TPCs must focus on explainability, a paradox still exists as to whether AI can be fully explained or not. Here we draw on the work of Sandra Wachter and colleagues at the Oxford Internet and Alan Turing Institutes in which they discuss the existence of a "right to explanation" in the European Union's General Data Protection Regulation (GDPR) program (https://gdpr.eu/what-is-gdpr/; GDPR.EU, 2022). According to the GDPR, the "data subject" must be provided with "meaningful information about the logic involved" in the automatic decision-making process, which is commonly referred to as the "right to explanation" (Wachter et al., 2018). Thus, an AI application is expected to provide accurate decisions and to justify these to end-users. Wachter et al. write:

> Attempts to implement a right to explanation that opens the 'black box' to provide insight into the internal decision-making process of algorithms face four major legal and technical barriers. First, a legally binding right to explanation does not exist in the GDPR. Second, even if legally binding, the right would only apply in limited cases (when a negative decision was solely automated and had legal or other similar significant effects). Third, explaining the functionality of complex algorithmic decision-making systems and their rationale in specific cases is a technically

challenging problem. Explanations may likewise offer little meaningful information to data subjects, raising questions about their value. Finally, data controllers have an interest in not sharing details of their algorithms to avoid disclosing trade secrets, violating the rights and freedoms of others (e.g. privacy), and allowing data subjects to game or manipulate the decision-making system.

(Wachter et al., 2018, pp. 3–4)

They emphasize that AI explanations can be pursued without opening the "black box," proposing three goals for explanations: "(1) to inform and help the subject understand why a particular decision was reached, (2) to provide grounds to contest adverse decisions, and (3) to understand what could be changed to receive a desired result in the future" (p. 4). They argue for using *counterfactuals* as a way to provide explanations. Essentially, the statement of an algorithmic decision is followed by a counterfactual, as in their example: "You were denied a loan because your annual income was £30,000. If your income had been £45,000, you would have been offered a loan" (p. 5). While XAI may typically refer to work to convey "the internal state or logic of an algorithm that leads to a decision… counterfactuals describe a dependency on the external facts that led to that decision" (p. 6). And while TPCs may have little ability to access and explain the "millions of variables intricately connected" in algorithms, TPCs are indeed skilled in providing explanations that convey information to describe a dependency that influences an outcome.

A vast amount of research continues to surround work to explain AI; for example, Ilia Stepin and colleagues (2021) surveyed both contrastive and counterfactual explanation methods for XAI, highlighting shortcomings with both approaches. At this point, we cannot completely open each algorithmic black box; we cannot make AI completely transparent for users. As Orhan Yalcin (2020) writes:

> AI systems have already become so complicated that it has become pretty much impossible for a regular user to understand how it works. I am sure that no more than %1 of Google Translate users know how it works, but we trust the system anyway and use it extensively. But, we should know how the system works, or at least, we should be able to get information about it when necessary.

In sum, TPCs must use other tactics. Industries are adopting AI; however, we have not yet adapted to it. Meanings will be made in ways that are different than we have done in the past. We clearly have evolving generations of decision-making and problem-solving machines; therefore, explainability and usability must likewise evolve to understand and promote trust.

Trustworthy AI

Cognitive systems researchers Peter Lewis and Stephen Marsh (2022), in their detailed functionalist perspective on trust in AI, question the efficacy of arguing either how to make people trust AI or how AI cannot be trusted as "people are going ahead and basically doing their own thing anyway, and… we should probably help them" (p. 33). They cite Bryson (2018) and Ryan (2020), who argue that no one should trust AI, in contrast then to Sutrop (2019), who describes trust as a valid notion when the AI has the capability to determine the best action for achieving a goal. While AI can be designed to be more trustworthy, Lewis and Marsh emphasize that this does not "compel anyone to actually trust it in any way." Rather, the trustworthiness of AI:

> is based on any number of traits, beliefs, desires, intentions, competencies, and so on.… One option for others to obtain evidence of our trustworthiness can come through observing the way we behave in some context.… Another option is by providing insight into how we work, think, and value, perhaps through transparency or explanations.
>
> *(p. 34)*

Again, TPCs are positioned to observe users, obtain and document evidence, and provide transparency and explanations.

Moreover, TPCs should be part of the ongoing development and use of guidelines for Trustworthy AI. The EU's *Ethics Guidelines for Trustworthy AI* (2019) propose that Trustworthy AI be lawful, ethical, and robust, both from a technical and social perspective. These guidelines propose the following requirements for Trustworthy AI: human agency and oversight; technical robustness and safety; privacy and data governance; transparency; diversity, non-discrimination, and fairness; environmental and societal wellbeing; and accountability. The companion Assessment List for Trustworthy AI (ALTAI, 2020) represents an initial approach to evaluating Trustworthy AI for the purpose of minimizing risks while maximizing the benefits of AI. It includes key questions for each guideline to encourage "thoughtful reflection to provoke appropriate action and nurture an organisational culture committed to developing and maintaining Trustworthy AI systems" (p. 4). We encourage TPCs to use ALTAI to evaluate Trustworthy AI. The first two guidelines include the following questions; note their relevance to TPC work:

- Human agency and oversight
  - Is the AI system designed to interact, guide or take decisions by human end-users that affect humans or society?
    - Could the AI system generate confusion for some or all end-users or subjects on whether a decision, content, advice or outcome is the result of an algorithmic decision?

- Are end-users or other subjects adequately made aware that a decision, content, advice or outcome is the result of an algorithmic decision?
- Have the humans (human-in-the-loop, human-on-the-loop, human-in-command) been given specific training on how to exercise oversight?
- Technical robustness and safety
  - Could the AI system have adversarial, critical or damaging effects (e.g. to human or societal safety) in case of risks or threats such as design or technical faults, defects, outages, attacks, misuse, inappropriate or malicious use?
  - Did you inform end-users of the duration of security coverage and updates?
  - Did you define risks, risk metrics and risk levels of the AI system in each specific use case?
  - Did you inform end-users and subjects of existing or potential risks? (pp. 7–10).

Also as a means to build Trustworthy AI development, Miles Brundage (2020), along with 58 worldwide co-authors, provides steps that different stakeholders in AI development can take to verify AI claims regarding the safety, security, fairness, and privacy protection of AI systems. The mechanisms they outline work to address these specific audience questions:

- Can I (as a user) verify the claims made about the level of privacy protection guaranteed by a new AI system I'd like to use for machine translation of sensitive documents?
- Can I (as a regulator) trace the steps that led to an accident caused by an autonomous vehicle? Against what standards should an autonomous vehicle company's safety claims be compared?
- Can I (as an academic) conduct impartial research on the impacts associated with large-scale AI systems when I lack the computing resources of industry?
- Can I (as an AI developer) verify that my competitors in a given area of AI development will follow best practices rather than cut corners to gain an advantage? (p. 1).

Ultimately, Lewis and Marsh (2022) ask:

> When and how will we decide whether to trust intelligent entities of various kinds, entities that we find ourselves in a position to interact with, both now and in the future?…
>
> We are thus faced with a simple choice: do we choose to disengage from reasoning using trust and trustworthiness for artefacts that appear

mind-like, when we cannot be fully sure they are capable of being trust-worthy? Or… do we choose to trust that this way of reasoning has the potential to be *useful*, in the context of systems that we cannot fully un-derstand, predict, or control?

(pp. 47–48)

Scholars across multiple disciplines have provided direction for understanding, examining, and designing AI-human interactions for the purpose of creating useful systems.

Implications for TPC Work

In 2018, innovative technical writer Tom Johnson wrote in his blog *I'd Rather Be Writing* about hybrid TPC roles:

When we try to sell our tech comm skills, promoting our *writing* skills doesn't seem to impress people anymore, as writing is considered more of a presumed skill everyone has. To give a sense of value, we need to hyphenate our job titles, becoming more of a hybrid professional.

He included a poll asking TPCs to indicate which role of 28 listed best de-scribed their work, to which 360 TPCs from around the world responded. The role technical writer/editor represented the largest percentage, with 14.12% of respondents, followed by technical writer/content developer, with 7.25%; technical writer/information architect, with 6.77%; technical writer/API docs specialist, with 4.87%; technical writer/doc tools guru, with 4.48%; technical writer/DITA specialist, with 3.34%; and technical writer/UX copywriter and technical writer/programmer, with 3.24% each. All other hybrid roles received less than 3%, ranging from technical writer/video producer, with 2.58%, to technical writer/web analytics and SEO maven, with 0.29%.

To examine writer identity, socio-technological literacies, and collaboration practices, Duin and Breuch (in press) conducted semi-structured interviews with 20 TPC leaders who are members of their programs' Technical Commu-nication Advisory Board. Results indicated a further broadening of TPC iden-tities. Abilities critical to the 2020 TPC workplace included working remotely, collaborating, thinking strategically, building relationships and networks, and expanding understanding of content authoring, tools and platforms, translation and localization, business ROI, legal and regulatory compliance, and usability/audience.

Again, the purpose of this book is to cultivate an even deeper understanding of human augmentation technology, to build TPC capacity to articulate its benefits and risks as part of current and future work and collaboration. TPC roles evolve with technological change, so we must consider how roles are likely to be augmented with AI or replaced by it along with new roles that will emerge. The TPC role has always been to *explain*; with augmentation technology, there is much to identify and explain. While engineers work on very specific augmentation technology development, e.g., specific facial expressions for a robot, TPCs build usability through work to identify the best facial expression for the specific audience and task. As TPCs write less, they monitor more.

All of the roles that Tom Johnson included in 2018 are now being augmented with AI; all programming, content development, information architecture, writer and producer, and usability roles increasingly rely on AI automation. TPCs who do not work to increase their AI literacy may well find their roles decreased.

Numerous studies predict that increasing percentages of the workforce will be replaced by AI/augmentation technologies over the coming years. Barry Dainton et al. (2021), in their introduction to the edited collection *Minding the Future*, assert the following:

> Predictions are of course risky, but the jobs widely believed to be most at risk include factory workers, lawyers, accountants and taxi drivers—and by the time GPT-5 arrives philosophers, poets and novelists might be at risk too. Working out ways of responding to these developments which maximize the potential benefits while minimizing unwanted disruption is likely [to] be among the greatest social and political challenges facing us over the next few decades.
>
> *(p. 10)*

Milad Mirbabaie and colleagues (2022), in their study to understand AI identity threat in the workplace, found three central predictors: changes to work, loss of position/status, and AI identity. One interesting finding from their extensive literature review was that individual experience with AI was shown to be less relevant than expected. They suggest that "by first explaining and defining AI identity and further revealing influencing factors on employees' identities in the workplace, practitioners should consider the special role of identity in the context of human–AI collaboration." Working to build understanding of digital and AI literacy, its competencies and design considerations, is one route toward explaining and defining AI identity. They also suggest introducing Explainable AI to address AI identity in the workplace. Therefore, roles that will increasingly emerge include technical writer/explainability specialist.

TPCs can also assist in helping users identify how their identities will change. Tori Homann (2022), writing for the company North American Signs, identifies automation and augmentation trends for 2021, writing that:

> Even though the thought of robots and AI taking over some tasks of the workforce can be scary, there is also a lot of growth potential.... By

removing the burden of repetitive or low-level problem-solving tasks, workers are freed up to contribute to value-adding tasks. So when a worker performs in tandem with automated technology, their job is augmented so that they can accomplish even more work in less time with fewer mistakes.

The Deloitte Corporation calls this worker plus automated technology a super-job; North American Signs calls it a superteam (Mallon et al., 2020). Most critical to individual and organizational success will be collective understanding of socio-ethical possibilities and consequences of augmentation technologies and AI, our focus in Chapter 5.

TPCs in user experience and support specialist roles will find their positions changing amid algorithmic AI technologies. One example is the need to position chatbots between front-line technicians and self-help or self-service as the chatbot provides an interface where users can interact with the information in a more natural way. The user receives direction; the chatbot learns from the user. TPCs will need to become more conscious of the differences in a chatbot's use of terms across rhetorical contexts. TPCs continue to evolve from solving problems for users to assisting people in solving problems themselves. Moreover, augmentation technologies and AI are increasingly reliant on conversations, those between humans and other humans, humans and machines, and machines with other machines. Numerous job postings exist for conversation design specialists who gather business needs and then develop text and media content for AI virtual agents to understand and interact with end-users.

In short, augmentation technologies and AI demand critical attention to multiple literacies including digital, data, and AI literacies. Digital literacy is about understanding and technological embodiment; data literacy is about understanding, decision-making, and protection (privacy, surveillance); and AI literacy is about understanding, explaining, and influencing the logic of AI systems. As we build these multiple literacies, TPC work continuously evolves to include that of technical writer/explainability specialist/chatbot trainer/conversation designer.

References

Adams, K. (2021). *31 numbers that show how big Epic, Cerner, Allscripts & Meditech are in healthcare.* Becker's Health IT. https://www.beckershospitalreview.com/healthcare-information-technology/31-numbers-that-show-how-big-epic-cerner-allscripts-meditech-are-in-healthcare.html

ALA (2022). *Digital literacy.* American Library Association. https://literacy.ala.org/digital-literacy/

Assessment List for Trustworthy Artificial Intelligence. (2020). European Union, High-Level Expert Group on Artificial Intelligence. https://futurium.ec.europa.eu/en/european-ai-alliance/pages/altai-assessment-list-trustworthy-artificial-intelligence

Barredo Arrieta, A., Díaz-Rodríguez, N., Del Ser, J., Bennetot, A., Tabik, S., Barbado, A., Garcia, S., Gil-Lopez, S., Molina, D., Benjamins, R., Chatila, R., & Herrera, F. (2020). Explainable Artificial Intelligence (XAI): Concepts, taxonomies, opportunities and challenges toward responsible AI. *Information Fusion, 58*, 82–115. https://doi.org/10.1016/j.inffus.2019.12.012

Bawden, D. (2001). Information and digital literacies: A review of concepts. *Journal of Documentation, 57*(2), 218–259.

Bawden, D. (2008). Origins and concepts of digital literacy. In C. Lankshear and M. Knobel (Eds.), *Digital literacies: Concepts, policies and practices* (pp. 17–32). Peter Lang.

Beck, E. N. (2015). The invisible digital identity: Assemblages in digital networks. *Computers and Composition, 35*, 125–140.

Beck, E., & Campos, L. H. (Eds.). (2020). *Privacy matters: Conversations about surveillance within and beyond the classroom.* University Press of Colorado. http://www.jstor.org/stable/j.ctv1gk4r7z

Bray, H. (2022, May 15). MIT, Harvard scientists find AI can recognize race from X-rays—and nobody knows how. *Boston Globe.* https://www.bostonglobe.com/2022/05/13/business/mit-harvard-scientists-find-ai-can-recognize-race-x-rays-nobody-knows-how/

Breuch, L. K. (2002). Thinking critically about technological literacy. *Technical Communication Quarterly, 11*(3), 267–288.

Brumberger, E. R., Lauer, C., & Northcut, K. (2013). Technological literacy in the visual communication classroom: Reconciling principles and practice for the "whole" communicator. *Programmatic Perspectives, 5*(2), 171–196.

Brundage, M. et al. (2020). *Improving verifiability in AI development.* https://openai.com/blog/improving-verifiability/

Bryson, J. (2018). *AI & global governance: No one should trust AI.* United Nations University Centre for Policy Research Article. https://cpr.unu.edu/publications/articles/ai-global-governance-no-one-should-trust-ai.html

Cargile Cook, K. (2002). Layered literacies: A theoretical frame for technical communication pedagogy. *Technical Communication Quarterly, 11*(1), 5–29.

Carretero Gomez, S., Vuorikari, R. & Punie, Y. (2017). *DigComp 2.1: The Digital Competence Framework for Citizens with eight proficiency levels and examples of use.* EUR 28558 EN. Publications Office of the European Union, Luxembourg. https://doi.org/10.2760/38842

Dainton, B., Slocombe, W., & Tanyi, A. (2021). *Minding the future: Artificial intelligence, philosophical visions and science fiction.* Springer.

Devlin, H. (2017, April 13). AI programs exhibit racial and gender biases, research reveals. *Guardian.* https://www.theguardian.com/technology/2017/apr/13/ai-programs-exhibit-racist-and-sexist-biases-research-reveals

Dickson, B. (2020). *What is artificial intelligence (Narrow AI)?* TechTalks. https://bdtechtalks.com/2020/04/09/what-is-narrow-artificial-intelligence-ani/

D'Ignazio, C., & Bhargava, R. (2015). *Approaches to building big data literacy.* Bloomberg Data for Good Exchange Conference, New York, September 28, 2015. https://dam-prod.media.mit.edu/x/2016/10/20/Edu_D%27Ignazio_52.pdf

Duin, A. H., & Breuch, L. K. (in press). Writer identity, literacy, and collaboration: 20 technical communication leaders in 2020. In L. Arduser (Ed.), *Workplace writing.* CSU Press TPC Foundations and Innovations series.

Duin, A. H., & Pedersen, I. (2023). *Competencies, design considerations, and new roles for work with augmentation technologies and AI* [Multimedia collection]. Fabric of Digital Life. https://fabricofdigitallife.com/Browse/objects/facets/collection:72

Duin, A. H., & Pedersen, I. (2021). *Writing futures: Collaborative, algorithmic, autonomous.* Springer.

Duin, A. H., Pedersen, I., & Tham, J. (2021). Building digital literacy through exploration and curation of emerging technologies: A networked learning collaborative. In N. B. Dohn, S. B. Hansen, J. J. Hansen, M. deLaat, & T. Ryberg (Eds.), *Conceptualizing and innovating education and work with networked learning* (pp. 93–114). Springer.

Duin, A. H., & Tham, J. (2018). Cultivating code literacy: A case study of course redesign through advisory board engagement. ACM SIGDOC, *Communication Design Quarterly, 6*(3), 44–58.

Educational Testing Service. (2003). *Succeeding in the 21st century. What higher education must do to address the gap in information and communication technology proficiencies. Assessing literacy for today and tomorrow.* Princeton, NJ: Educational Testing Service.

European Commission (2022, April 26). *Ethics guidelines for trustworthy AI.* https://digital-strategy.ec.europa.eu/en/library/ethics-guidelines-trustworthy-ai

Ferreira, J. J., & Monteiro, M. (2021). Designer-user communication for XAI: An epistemological approach to discuss XAI design. ACM CHI Workshop on Operationalizing Human-Centered Perspectives in Explainable AI at CHI 2021. https://doi.org/10.48550/arXiv.2105.07804

Fjeld, J., Achten, N., Hilligoss, H., Nagy, A., & Srikumar, M. (2020). *Principled artificial intelligence: Mapping consensus in ethical and rights-based approaches to principles for AI.* Berkman Klein Center Research Publication. https://ssrn.com/abstract=3518482

GDPR.EU (2022). *European Union's General Data Protection Regulation (GDPR) program.* https://gdpr.eu/

Gichoya, J. W., et al. (2022). AI recognition of patient race in medical imaging: A modelling study. *Lancet Digit Health, 4*, e406–e414. https://doi.org/10.1016/S2589-7500(22)00063-2

Gilster, P. (1997). *Digital literacy.* John Wiley & Sons.

Global Digital Literacy Council (2022). *Signing and ratification of the global standard 6.* https://www.gdlcouncil.org/

Hancock, J. T., Naaman, M., & Levy, K. (2020). AI-mediated communication: Definition, research agenda, and ethical considerations. *Journal of Computer-Mediated Communication, 25*(1), 89–100. https://doi.org/10.1093/jcmc/zmz022

Harrod, J. (2020). *AI literacy, or why understanding AI will help you every day.* https://fabricofdigitallife.com/Detail/objects/4847

He, J., Baxter, S. L., Xu, J., Xu, J. Zhou, X., & Zhang, K. (2019). The practical implementation of artificial intelligence technologies in medicine. *Nature Medicine, 25*(1), 30–36. https://dx.doi.org/10.1038/s41591-018-0307-0

Heath, N. (2018). *What is artificial general intelligence?* ZDNet/TechRepublic. https://www.zdnet.com/article/what-is-artificial-general-intelligence/

Homann, T. (2022). *Automation & augmentation: Trends for 2021.* North American Signs. https://www.northamericansigns.com/automation-and-augmentation/

Hovde, M.R. & Renguette, C.C. (2017). Technological literacy. *Technical Communication Quarterly, 26*(4), 395–411.

IBM (2022). *Explainable AI.* https://www.ibm.com/watson/explainable-ai

Inam, R., Terra, A., Mujumdar, A., Fersman, E., & Feljan, A. V. (2022). *Explainable AI—how humans can trust AI.* Ericsson.com. https://www.ericsson.com/en/reports-and-papers/white-papers/explainable-ai--how-humans-can-trust-ai

International Society for Technology in Education. (2016). *ISTE standards for students.* https://www.iste.org/standards/for-students

Jacobs, G., & Castek, J. (2018). Digital problem solving: The literacies of navigating life in the digital age. *Journal of Adolescent and Adult Literacy, 61*(6), 681–685.

Jin, W., Fan, J., Gromala, D., Pasquier, P., & Hamarneh, G. (2021). *EUCA: The End-User-Centered explainable AI framework*. Cornell University. https://doi.org/10.48550/arXiv.2102.02437

Jisc (2022). *Building digital capability*. https://www.jisc.ac.uk/building-digital-capability

Johnson, T. (2018). If writing is no longer a marketable skill, what is? [Blog post]. *I'd Rather Be Writing*. https://idratherbewriting.com/2018/08/09/writing-no-longer-a-skill/

Kanade, V. (2022). *What is general artificial intelligence (AI)? Definition, challenges, and trends*. Toolbox Tech. https://www.toolbox.com/tech/artificial-intelligence/articles/what-is-general-ai/

Koenig, A. (2020). The algorithms know me and I know them: Using student journals to uncover algorithmic literacy awareness. *Computers and Composition, 58*.

Lanham, R. (1995). Digital literacy. *Scientific American, 273*(3), 160–161.

Lankshear, C., & Knobel, M. (Eds.). (2008). *Digital literacies–concepts, policies and practices*. Peter Lang.

Lepore, J. (2020a). *If then: How the Simulmatics Corporation invented the future*. Liveright.

Lepore, J. (2020b). Scientists use big data to sway elections and predict riots—welcome to the 1960s. *Nature, 585*, 348–350. https://doi.org/10.1038/d41586-020-02607-8

Lewis, P. R., & Marsh, S. (2022). What is it like to trust a rock? A functionalist perspective on trust and trustworthiness in artificial intelligence. *Cognitive Systems Research, 72*, 33–49. https://doi.org/10.1016/j.cogsys.2021.11.001

Library Study Smart, Western Sydney University (2022). *What is digital literacy?* https://www.westernsydney.edu.au/studysmart/home/study_skills_guides/digital_literacy/what_is_digital_literacy

Long, D., & Magerko, B. (2020). *What is AI literacy? Competencies and design considerations*. CHI '20, April 25–30, 2020, Honolulu, HI, USA. https://doi.org/10.1145/3313831.3376727

Mallon, D., Van Durme, Y., Hauptmann, M., Yan, R., & Poynton, S. (2020). *Superteams: Putting AI in the group*. Deloitte Insights. https://www2.deloitte.com/us/en/insights/focus/human-capital-trends/2020/human-ai-collaboration.html

Mirbabaie, M., Brunder, F., Mollmann, N. R. J., & Stieglitz, S. (2022). The rise of artificial intelligence—understanding the AI identity threat at the workplace. *Electronic Markets, 32*, 73–99. https://doi.org/10.1007/s12525-021-00496-x

Museum and Library Services Act of 2010. (2010). Institute of Museum and Library Services. https://www.imls.gov/sites/default/files/mlsa_2010_2.pdf

Northcut, K. M., & Brumberger, E. R. (2010). Resisting the lure of technology-driven design: Pedagogical approaches to visual communication. *Journal of Technical Writing and Communication, 40*(4), 459–471.

Oracle advertising. (2022). Oracle. https://www.oracle.com/cx/advertising/

Panetta, K. (2021). *A data and analytics leader's guide to data literacy*. Gartner. https://www.gartner.com/smarterwithgartner/a-data-and-analytics-leaders-guide-to-data-literacy

Pangrazio, L., & Selwyn, N. (2018). 'Personal data literacies': A critical literacies approach to enhancing understandings of personal digital data. *New Media & Society, 21*(5). http://dx.doi.org/10.1177/1461444818799523

Raden, N. (2021). *How did a proprietary AI get into hundreds of hospitals—without extensive peer reviews? The concerning story of Epic's Deterioration Index*. Diginomica. https://diginomica.com/how-did-proprietary-ai-get-hundreds-hospitals-without-extensive-peer-reviews-concerning-story-epics

Ribera, M., & Lapedriza, A. (2019). Can we do better explanations? A proposal of user-centered explainable AI. In *Joint Proceedings of the ACM IUI 2019 Workshops*, March 20, 2019, Los Angeles, CA, USA.

Romero, A. (2021). *GPT-4 will have 100 trillion parameters—500x the size of GPT-3.* Towards Data Science. https://towardsdatascience.com/gpt-4-will-have-100-trillion-parameters-500x-the-size-of-gpt-3-582b98d82253

Ryan, M. (2020). In AI we trust: Ethics, artificial intelligence, and reliability. *Science and Engineering Ethics, 26*, 2749–2767.

Selber, S. (2004). *Multiliteracies for a digital age.* Carbondale, IL: Southern Illinois U. Press.

Stepin, I., Alonso, J.M., Catala, A., & Pereira-Farina, M. (2021). A survey of contrastive and counterfactual explanation generation methods for explainable artificial intelligence. *IEEE Access*, 11974–12000.

Stordy, P. (2015). Taxonomy of literacies. *Journal of Documentation, 71*(3), 456–476. https://doi.org/10.1108/JD-10-2013-0128

Sutrop, M. (2019). Should we trust artificial intelligence? *Trames, 23*(4), 499–522.

Traxler, J. (2018). Digital literacy. *Research in Learning Technology, 26*, 1–21.

Turek, M. (2018). *Explainable artificial intelligence (XAI).* Defense Advanced Research Projects Agency. https://www.darpa.mil/program/explainable-artificial-intelligence

Turnley, M. (2007). Integrating critical approaches to technology and service-learning projects. *Technical Communication Quarterly, 16*(1), 103–123.

University of Helsinki (2022). *The Finnish 'elements of AI' online course trains employees of the European Union* [Press release]. https://www.helsinki.fi/en/news/people-and-technology/finnish-elements-ai-online-course-trains-employees-european-union

University of Minnesota (2022, October 1). *University of Minnesota develops AI algorithm to analyze chest X-rays for COVID-19.* Epic.com. https://www.epic.com/epic/post/university-minnesota-develops-ai-algorithm-analyze-chest-x-rays-covid-19

Vanek, J. (2020). *Issue brief: Digital literacy.* American Institutes for Research. https://www.air.org/sites/default/files/TSTMDigitalLiteracyBrief-508.pdf

Vanek, J. (2017). *Using the PIAAC framework for problem solving in technology-rich environments to guide instruction: An introduction for adult educators.* https://static1.squarespace.com/static/51bb74b8e4b0139570ddf020/t/589a3d3c1e5b6cd7b42cddcb/1486503229769/PSTRE_Guide_Vanek_2017.pdf

Varhol, P. (2021). *Why software testing needs explainable AI.* TechBeacon. https://techbeacon.com/app-dev-testing/why-software-testing-needs-explainable-ai

Verhulsdonck, G., & Tham, J. (2022). Tactical (dis)connection in smart cities: Post-connectivist technical communication for a datafied world. *Technical Communication Quarterly.* https://doi.org/10.1080/10572252.2021.2024606

Wachter, S., Mittelstadt, B., & Russell, C. (2018). Counterfactual explanations without opening the black box: Automated decisions and the GDPR. *Harvard Journal of Law & Technology, 31*, 1–52.

WIPO (2019). *WIPO technology trends 2019: Artificial intelligence.* World Intellectual Property Organization. https://www.wipo.int/edocs/pubdocs/en/wipo_pub_1055.pdf

Yalcin, O. G. (2020). *5 significant reasons why explainable AI is an existential need for humanity.* Towards Data Science. https://towardsdatascience.com/5-significant-reasons-why-explainable-ai-is-an-existential-need-for-humanity-abe57ced4541

5

SOCIO-ETHICAL CONSEQUENCES AND DESIGN FUTURES

Overview

Chapter 5 addresses potential socio-ethical consequences of augmentation technologies and artificial intelligence (AI). It provides clarity on biometric data and its relevance to augmentation technologies in combination with AI. *Principled Artificial Intelligence: Mapping Consensus in Ethical and Rights-Based Approaches to Principles for AI* (Fjeld et al., 2020) provides the ethical focus. The chapter uses work published by the AI Now Institute, which monitors and reports on human rights issues through several published reports. These include focus on algorithmic mining and bias, racial and gender discrimination, ableism, digital divides, unethical AI practices, misinformation, and other socio-ethical harms to humans. The chapter builds rationale for understanding digital and AI literacy as a means to avoid socio-ethical harms to humans that occur when digital and AI literacy is absent. The chapter discusses socio-ethical themes drawn from AI ethical domains and introduces principles that fall under them.

Key Questions

- How should technical and professional communication (TPC) professionals prepare for the new challenges that arise with augmentation technologies that use embodied technology, biometric recognition, and AI decision-making systems in combination?
- How can TPC professionals help mitigate unintended socio-ethical consequences, including algorithmic bias, racial and gender discrimination, ableism, digital divides, unethical practices, unsafe or dangerous situations, and other socio-ethical harms to humans?

DOI: 10.4324/9781003288008-7

- Universal human rights play a fundamental role in the definitions, associated concepts, and AI principles. How can we ensure they inform design, adoption, and adaptation to augmentation technologies?

Chapter 5 Links

Throughout the chapter, we refer to articles, videos, and reports which can be found in a related chapter collection at Fabric of Digital Life called *Socio-ethical consequences and design futures* (Duin & Pedersen, 2023). You can find a link to this collection at https://fabricofdigitallife.com.

Introduction

Meta's Reality Labs is publicly sharing news about developing highly immersive XR systems that will enhance the senses. In one video, the idea of "being in constant conversation with the machine" is used to describe both the extent of visual, aural, and haptic interfaces and the state of continuous immersion. Sean Keller, Director of Research Science at Reality Labs, predicts the movement from "personal computing to personalized computing" (Tech at Meta, 2021). It means designing a system for the combination of one's real-world experiences and virtual world activities beyond current augmented or virtual reality systems. The result could bring about a significant turn to human-centered computing with transformative results for people in their everyday lives.

Yet, these pre-release announcements do not discuss how these enhancement interfaces will extract, process, transmit, or store the biometric data needed to run these systems. They do not discuss if they were tested for different demographic groups, nor do they express harms that might arise. Because they are advertising, they do not discuss *potential* safety issues, marginalization of individuals in social virtual reality, or manipulation by external actors. They don't question the effect on race, gender, or intersectional identity. In Chapter 1, we discussed how worldview perspectives on augmentation technologies become popularized within the public sphere. Popular discourses can obfuscate the critical assessment of potential future harms. In the context of AI technology emergence, the *Disability, Bias, and AI* report explains:

> Many significant ethical choices are already made by the time AI is delivered to (or used on) consumers, meaning that market forces—not the needs of individuals and communities on whom AI acts—very often determine both which products are offered, and the limits and implications of these products.
>
> *(Whittaker et al., 2019, p. 7)*

While TPC scholarship generally attends to privacy, accountability, safety, and transparency, we need additional vigilance regarding augmentation technologies.

We reiterate our definition of augmentation technologies: they enhance human capabilities or productivity by adding to the body (or ambient environment around the body) cognitive, physical, sensory, and/or emotional enhancements. Consequently, the grounds on which TPC professionals work when communicating details about these data sources and infrastructures move into much more personal user spaces, and personal digital contexts, than ever before. Human control of technology, fairness and non-discrimination, professional responsibility—values that we expect to uphold—could become significantly compromised as users adopt and adapt to these cutting-edge technologies.

Augmentation technologies are embodied technologies that often use biometric recognition. Biometrics "is the science of recognizing the identity of a person based on the physical or behavioral attributes of the individual such as face, fingerprints, voice and iris" (Jain et al. 2010, p. vii). We draw on the AI Now Institute's *Regulating Biometrics: Global Approaches and Urgent Questions* report (Kak, 2020) to inform the scholarly landscape of issues relevant to biometrics as well as how they are regulated. There are numerous definitions of biometric data, however, one proposed by Els Kindt (2020) from the report provides a specific definition geared more directly to human beings. Biometric data can be defined as:

> all personal data (a) relating directly or indirectly to unique or distinctive biological or behavioural characteristics of human beings and (b) used or fit for use by automated means (c) for purposes of identification, identity verification, or verification of a claim of living natural persons.
>
> *(p. 62)*

Kindt's definition focuses not only on a description of personal biometric data, but also on the further purposes for using them by external actors such as automation.

Our book uses international, multi-stakeholder documents as much as possible given the global nature of the topic. In Chapter 1, we turned to a report published by Harvard University's Berkman Klein Center called *Principled Artificial Intelligence: Mapping Consensus in Ethical and Rights-based Approaches to Principles for AI* (Fjeld et al., 2020). It analyzed the contents of 36 prominent international AI principles documents, identifying ethical themes. These documents vary in intended audience and scope:

> they come from Latin America, East and South Asia, the Middle East, North America, and Europe, and cultural differences doubtless impact their

contents. Perhaps most saliently, though, they are authored by different actors: governments and intergovernmental organizations, companies, professional associations, advocacy groups, and multi-stakeholder initiatives.

(Fjeld et al., 2020, p. 4)

The result is the list of ethical themes: Promotion of Human Values, Professional Responsibility, Human Control of Technology, Fairness and Non-Discrimination, Transparency and Explainability, Safety and Security, Accountability, and Privacy.

We draw from this document because it strives for consensus. With the dizzying number of AI principles documents, AI ethical frameworks, drafted regulations, and corporate guidance documents, this report provides clarity without minimizing or erasing the voices of relevant stakeholders. It works to overcome the synonym problems in this field by analyzing the intent in certain passages. For example, stakeholders might be discussing *security* issues by using other terminology such as *protection*; this document's commitment to consensus helps to ameliorate misunderstandings. Finally, *Principled Artificial Intelligence* mediates the complex socio-technical assemblages with cross-references to overlapping issues.

In the next section, we explore why risks for augmentation technologies are novel.

Different Risks with Augmentation Technologies

In their article "Überveillance and the Rise of Last-Mile Implantables," Katina Michael et al. (2020) write about pervasive technologies. Like augmentation technologies, "pervasive technologies violate all of the aforementioned borders of privacy and so society must enter into the debate to address these issues. History has shown that commonly, consequences are delayed; false senses of security often exist in early stages" (p. 119). They provide a list of "Six Principal Risks of Pervasive Implantable Technologies" that breaks down issues into clear concepts and simplified questions relevant to Augmentation:

1   Insightfulness: With data gleaned across all veillances, devices will assess humans in multiple contexts, capacities, and times, allowing the system to have a precise and profound understanding of a human in their past, present, and future states.
2   Imperceptibility: Users will be mostly unaware of what is collected, by whom, for how long, how it is synthesized with other data, and who owns the data.
3   Incomprehensibility: Terms and conditions are often murky and/or mutable, and the everyday consumer is not likely to comprehend the

wide-ranging system, nor the associated risks across multiple organizations sharing data.

4   Indelibility: Our digital footprints are likely to leave an indelible history of analyzable behaviors, especially if we do not own our data, or if it is shared and stored elsewhere in the veillances.

5   Invasiveness: As we allow devices to listen inside of us and communicate back and forth between the veillances, we are likely to create systems in which not only are our behaviors predicted but even our intent. Dignity is likely to be at risk, even if unintentionally.

6   Involuntariness: Opting in to technology is becoming a requirement to participate in society, to belong and benefit socially or financially (Michael et al., 2020, pp. 118–119).

Augmentation technologies likewise permit external actors to have "precise and profound" knowledge about our behaviors, which in turn justifies predictions that are used to evaluate us and our intentions. Similarly, the notion that these assessments are completely voluntary is questionable. Users are asked to join and participate on third party platforms for work without necessarily being aware of their personal privacy rights.

Dataveillance is not a new concept (Andrejevic, 2007; van Dijk, 2014), but what is new is the advanced proficiency of augmentation technologies to perform surveillance activities. Andrew Iliadis researches ingestible technology and explains that computerized embodiment takes place "not [in] the exterior world but rather [through] the enhanced activity of the user's body—the user does not necessarily *interact* with the environment but *becomes* the environment" (2020, p. 6). Novel, also, are the sophisticated data infrastructures proposed to host biotechnologies by private sector companies. Safety and security changes when interfaces surround bodies and data extraction is imperceptible (Michael et al., 2020; Pedersen, 2020).

Mohamed Khamis and Florian Alt (2021) explain:

> communicating biometric data to surrounding devices or to remote servers may introduce a new attack channel: if data is sent in an unencrypted way, man-in-the-middle attacks can result in attackers intercepting sensitive information about the user as they are transferred wirelessly. This can happen without the user's control
>
> *(pp. 265–266)*

However, extensive AI hype obfuscates public inquiry about them. Katina Michael and her colleagues explain that knowledge about these systems is murky, suggesting that the risks are equally incomprehensible (Michael et al., 2020). The result is that citizens are bound to asymmetrical power dynamics as users adopt and adapt to these technologies. We promoted AI explainability (making sure that systems are capable of being understood) and AI literacy (making sure that citizens can understand how they are used) in Chapter 4, and we discuss them later in this chapter.

In the next section, we unpack each ethical theme and describe the related AI principles, re-envisioned for augmentation technologies. We discuss each theme and introduce principles that fall under them. We also explore relevant consequences for specific augmentation technologies and point to future designs and practices that may prevent or mitigate harms.

Ethical Themes, Principles, and Consequences for Augmentation Technology

We ask that our readers consider the ways that principles and implications overlap in complex ways. For instance, a discriminatory outcome of using an augmentation technology (a consequence) will need to be rectified by considering Fairness and Non-discrimination (a socio-ethical theme), but it may also require consideration of principles falling under Privacy (e.g., consent to use data), Safety and Security (e.g., public trust that AI systems have not been controlled by malicious actors), and Professional Responsibility (e.g., that the system has been designed responsibly).

Below, we itemize each theme and list the many principles that fall under each theme (e.g., the principle of Consent falls under the theme of Privacy). For each, we provide a short, basic principle definition to provide better clarity.

Theme 1: Privacy

Privacy is likely the most prevalent theme in any discussion of ethics and digital technology; AI technologies have only increased people's concerns over the loss of personal privacy. A central reason for this concern is agency and whether people have it when these technologies are deployed. The other reason is the constant extraction of personal data from users (data providers) for platforms, without the ability for users to retain control of that data. Without agency or control, one fears many outcomes at the hands of malicious actors. Finally, AI technologies and subsequent augmentation technology users are contextualized by the privacy paradox. The privacy paradox is the dichotomy between people's value-based concern to protect their personal privacy and their actual behavior in digital marketplaces, disclosing personal information.

Augmentation technologies involve exchanging biometric data in a manner that challenges privacy principles. "If we digitize our senses and cognitive processes, what kind of privacy intrusions do we potentially expose ourselves to?" (Dingler et al., 2021, p. 6). Biometric recognition technologies funnel bodily data to external actors. According to Amba Kak and drawing on the work of Aaron K. Martin and Edgar A. Whitley (2013), biometric recognition technologies "describe systems that 'fix' official identities to bodily, physiological, or behavioral traits, providing new ways for individuals to identify themselves, and also to be identified or tracked" (Kak, 2020, p. 6.). This kind of identification adds further vulnerability in that "beyond [simply] identifying people, these systems increasingly claim to be able to infer demographic characteristics, emotional states, and personality traits from bodily data" (Kak, 2020, p. 6).

To value privacy means to uphold that "AI systems should respect individuals' privacy, both in the use of data for the development of technological systems and by providing impacted people with agency over their data and decisions made with it" (Fjeld et al., 2020). Privacy as a theme breaks down into several key and much more specific principles: "Consent," "Control over the Use of Data," "Ability to Restrict Processing," "Right to Rectification," "Right to Erasure," "Privacy by Design," and "Recommend[ation for] Data Protection Laws" (Fjeld et al., 2020, p. 21). In Chapter 4, we discussed digital literacy, data literacy, and AI literacy as required competencies. Data literacy covers the ability to read, work, analyze, and argue with data as well as to understand and protect one's personal data.

Focus Agent

To help us further unpack privacy as an abstract concept, let's explore the proposal for a specialized conversational agent, an augmentation technology, in research and development at Microsoft. The focus agent is part of the HUE: Human Understanding and Empathy group at Microsoft. The description below was drawn from an article called "Building an AI that Feels." It demonstrates how expectations of personal privacy might be challenged because of the extensive biofeedback data extracted. Named a "focus agent," it bears the mark of augmentation technology in its technical description:

> Our 'focus agent' aimed to boost productivity by helping users schedule time to work on important tasks and helping them adhere to their plans. A camera (1) and computer software (2) kept track of the user's behavior. The sensing framework (3) detected the number of people in view and the user's position in front of the computer screen, estimated the user's emotional state, and also kept track of the user's activity within various applications. The agent app (4) controlled the focus agent avatar that engaged the user in conversation, using an AI-powered conversation bot (5) that drew on a variety of dialogue models to respond to the situation as appropriate.
> *(Czerwinski et al., 2021)*

The focus agent is deliberately distinguished as having "emotional intelligence" because it would so thoroughly sense users' biofeedback (Czerwinski et al.,

2021). It uses affective computing to encourage better workplace productivity, essentially to make people more efficient.

As we have discussed in earlier chapters, the intent to assess emotional responses to boost productivity—to create what one might describe as an *augmented worker*—would be classified under emotional enhancement. Related research by Grover and colleagues (2020) further describes the data extraction conducted through multiple data streams:

> Sensing Software: We leveraged a desktop sensing framework application that uses audio and video input (from a speaker and webcam) to continually detect the user's expressed emotional state…. The application uses a trained machine learning model to detect changes in the user's facial expression with high temporal resolution (multiple times per second). The sensing software was used to detect multiple different data streams from the users simultaneously, including the number of faces detected in the webcam view, and measures of the user's facial expression in different categories: neutral, anger, contempt, surprise, disgust, sadness, and fear using convolutional neural networks trained on handlabeled images.
>
> *(Grover et al., 2020, p. 394)*

In a nutshell, a worker's highly personal reactions become contextualized as a measure of their work.

The focus agent example is relevant to many aspects of this book, but the key privacy issue surrounds the monitoring of human behavior through biometric data extraction, measurement, and assessment. The focus agent involves workers being continually monitored by webcams to determine their "expressed emotional state" (Grover et al., 2020, p. 394). Emotions are detected through facial expressions, a practice that has been challenged by previous researchers as inaccurate (Crawford, 2021). Such a condition of surveillance might make workers feel uneasy. Focus agent technology is a form of "platform biometrics," which Jeremy Crampton (2019) questions:

> we can understand the huge leaps that have taken surveillance beyond simply seeing and recognizing, to categorizing and inferring a subject's innermost nature…. In its review of many of these systems, the AI Now Institute warned of the dangers these systems present to mass surveillance due to its faulty emotion and affect recognition, its lack of scientific evidence, and its irresponsible ethics.

Questions to consider in TPC contexts:

• How are users asked (or compelled) to give consent to the usage of the biometric data, if at all?

- How is this biometric data extracted, handled, and manipulated?
- How is data stored? Can it be erased at the request of data subjects?
- How are decision-making systems used to evaluate behaviors?
- And more conceptually, should citizens and workers divulge their emotional states for an employer? Should it be a requirement or even an option? Just because augmentation technologies *can* incent workers to be more productive through emotional monitoring, does that activity not breach workers' rights to privacy?

The focus agent inventors do make note of privacy issues as a recommendation for future work:

> Our work has a few important limitations that should be noted. First, our agent prototypes, particularly the VA [Virtual Agent] prototype, used users' webcams to continuously monitor their emotional state. User surveillance like this may result in some users feeling anxious that they are being monitored. These challenges are important and relevant, and we recommend future work to investigate the nature of user privacy concerns in this context.
>
> *(Grover et al,. 2020, p. 399)*

This "design stage" prototype reveals the potential negative human consequences but does not consider it a necessary part of the prototype. In contrast, we feel privacy concerns and potential consequences *must* inform the early design of augmentation technology. Mohamed Khamis and Florian Alt provide a framework for privacy and security assessment of augmentation technologies for designers under three categories: "Data Handling," "Awareness and Consent," and "Control over the Data." TPCs should be aware of these kinds of assessment tools, as they may be asked to write the supporting instructions for work-based monitoring apps.

With more workers at home due to the COVID pandemic, there is more of a (perceived) need to automate worker efficiency and perhaps see the justification for continuous monitoring of workers' emotional states. Pew Research reports that "roughly six-in-ten U.S. workers who say their jobs can mainly be done from home (59%) are working from home all or most of the time" (Parker et al., 2022). Is it ethical to impose this kind of monitoring on people who are at home and might be simultaneously dealing with pressing or even serious family issues?

Privacy Principles

TPC professionals need to be familiar with these principles because privacy issues will become implicated in development cycles for augmentation technology:

- **Consent** is the notion that "a person's data should not be used without their knowledge and permission. Informed consent is a closely related but more robust principle—derived from the medical field—which requires individuals be informed of risks, benefits, and alternatives" (Fjeld et al., 2020, p. 22).
- **Control over the Use of Data** is the notion that "data subjects should have some degree of influence over how and why information about them is used" (Fjeld et al., 2020, p. 22).
- **Ability to Restrict Processing** is the "power of data subjects to have their data restricted from use in connection with AI technology" (Fjeld et al., 2020, p. 23).
- **Right to Rectification** refers to the "right of data subjects to amend or modify information held by a data controller if it is incorrect or incomplete" (Fjeld et al., 2020, p. 24).
- **Right to Erasure** refers to "an enforceable right of data subjects to the removal of their personal data" (Fjeld et al., 2020, p. 24).
- **Privacy by Design** is "an obligation on AI developers and operators to integrate considerations of data privacy into the construction of an AI system and the overall lifecycle of the data" (Fjeld et al., 2020, p. 25).
- **Recommendation for Data Protection Laws** is the belief that "new government regulation is a necessary component of protecting privacy in the face of AI technologies" (Fjeld et al., 2020, p. 25).

Theme 2: Accountability

One highly publicized case concerning augmentation technology is the Second Sight retinal implant system, which essentially abandoned users when the "company had abandoned the technology and was on the verge of going bankrupt" (Strickland & Harris, 2022). Second Sight involves several combined innovations dating back more than a decade that promised degrees of sight restoration to blind persons and those with low vision. Using a retinal prosthesis system, "bionic eyes" and brain implants, Second Sight widely publicized its sensory enhancing technology emphasizing its augmentative capacity. Yet, in 2022, the company decided to stop supporting the technology. Fabric houses several videos and articles under the marketing keyword "Second Sight" to chart the rise and fall of this sensory enhancement augmentation technology. The story spans heavily hyped early innovation videos in 2012 to mainstream journalism that relayed the abandonment of the technology by the company in 2022. Eliza Strickland and Mark Harris's article "Their Bionic Eyes Are Now Obsolete and Unsupported" explains the impact:

> More than 350 other blind people around the world with Second Sight's implants in their eyes, find themselves in a world in which the technology

that transformed their lives is just another obsolete gadget. One technical hiccup, one broken wire, and they lose their artificial vision, possibly forever. To add injury to insult: A defunct Argus system in the eye could cause medical complications or interfere with procedures such as MRI scans, and it could be painful or expensive to remove.

(Strickland & Harris, 2022)

This article provides an in-depth look at all the actors involved in bringing the technology to the public, but also the devastating consequence of not putting accountability practices into place to protect these early adopters.

Accountability includes AI ethical principles "concerning the importance of mechanisms to ensure that accountability for the impacts of AI systems is appropriately distributed, and that adequate remedies are provided" (Fjeld et al., 2020, p. 5). Accountability for augmentation technologies hinges on regulating both AI systems and biometric data. Woodrow Hartzog (2020) explains, "Peoples' unique biometric identifiers, now easily wholesale collected and stored, are not like other kinds of authenticators like passwords and social security numbers because if they are compromised, they cannot be changed" (p. 99). Amba Kak writes:

> as legislation moves beyond traditional data privacy and security concerns to questions of accountability around whether or how to use these systems, and who is liable if these systems fail, some recent bills shift the focus from 'data' to 'systems.'
>
> *(Kak, 2020, p. 20)*

Accountability Principles

The Accountability theme can be mapped to three essential stages for AI systems that are relevant for augmentation technologies, "design (pre-deployment), monitoring (during deployment), and redress (after harm has occurred)" (Fjeld et al., 2020, p. 29). The list below illustrates the principles organized in each stage:

Stages of Design

- **Verifiability and Replicability** is the belief that "an AI experiment ought to 'exhibit the same behavior when repeated under the same conditions' and provide sufficient detail about its operations that it may be validated" (Fjeld et al., 2020, p. 29).
- **Impact Assessment** is a process that "captures both specific calls for human rights impact assessments (HRIAs) as well as more general calls for the advance identification, prevention, and mitigation of negative impacts of AI technology" (Fjeld et al., 2020, p. 30).

- **Environmental Responsibility** is a belief that "reflects the growing recognition that AI, as a part of our human future, will necessarily interact with environmental concerns, and that those who build and implement AI technology must be accountable for its ecological impacts" (Fjeld et al., 2020, p. 31).

Stages of Monitoring

- **Evaluation and Auditing Requirements** is the articulation of:

  the importance of not only building technologies that are capable of being audited, but also to use the learnings from evaluations to feed back into a system and to ensure that it is continually improved, 'tuning AI models periodically to cater for changes to data and/or models over time'.
  (Fjeld et al., 2020, p. 31)

- **Creation of a Monitoring Body** involves the realization that a "new organization or structure may be required to create and oversee standards and best practices in the context of AI" (Fjeld et al., 2020, p. 32).
- **Ability to Appeal** is "the possibility that an individual who is the subject of a decision made by an AI could challenge that decision" (Fjeld et al., 2020, p. 32).

Stages of Redress

- **Remedy for Automated Decision** means that "as AI technology is deployed in increasingly critical contexts, its decisions will have real consequences, and that remedies should be available just as they are for the consequences of human actions" (Fjeld et al., 2020, p. 33).
- **Liability and Legal Responsibility** means that we need to "ensure that the individuals or entities at fault for harm caused by an AI system can be held accountable" (Fjeld et al., 2020, pp. 33–34).
- **Recommends Adoption of New Regulations** means "AI technology represents a significant enough departure from the status quo that new regulatory regimes are required to ensure it is built and implemented in an ethical and rights-respecting manner" (Fjeld et al., 2020, p. 34).

Theme 3: Safety and Security

Technology advancements can place users in vulnerable scenarios before safety and security has been appropriately considered. Augmentation technologies augment highly personal activities and capabilities; however, they could simultaneously make people unsafe through those same points of contact. In a previous publication on embodied technologies, author Isabel Pedersen wrote about an event in 2017 when malicious actors created and released a computer virus to target all Bluetooth networks (Pedersen, 2020). Called BlueBorne, it allowed "attackers to take full control of devices... and spread malware laterally to adjacent devices" (Armis, 2017). The virus spread across "Bluetooth devices—smartphones, watches, tablets, speakers, fitness trackers, and even smart cars—searching for weak spots in these personal networks" (Pedersen, 2020, p. 21). It was later assessed as having the ability to "affect 8.2 billion units, the entire range of Bluetooth-enabled devices.... [Because it was] designed to be nearly invisible, the virus was developed to be infectious as it moved through the air, device to device over Wi-Fi with agility" (Pedersen, 2020, p. 21). BlueBorne is still a threat, especially in cases when devices are left on at all times, which is typical (Cameron, 2021). The point is that while wireless connectivity standards for personal devices have enabled people to become more physically safe, more socially and professionally connected, and more secure, they have simultaneously created an inherent condition of potential vulnerability across these connected devices, a safety paradox.

Harassment

Harassment in extended reality (XR) spaces has been highlighted as a serious safety issue, especially in social virtual reality (Blackwell et al., 2019). Several journalists have been following harassment in the Metaverse, including Sheera Frenkel and Kellen Browning in their *New York Times* article "The Metaverse's Dark Side: Here Come Harassment and Assaults" (2021). The British Broadcasting Corporation also covered it with an article, "Female Avatar Sexually Assaulted in Meta VR Platform, Campaigners Say" (2022). Lindsay Blackwell and her collaborators have written about VR, commenting on both harms such as harassment and affordances:

> We also find that the embodiment and presence afforded by VR can make harassment feel more intense; however, a few participants felt that embodiment and presence could reduce the incidence of harassment experiences by increasing empathy for other users, a finding supported by other empirical work.
>
> *(p. 2)*

Harassment behaviors take many forms:

> Participants' specific experiences of harassment in social VR largely fell into three categories: *verbal harassment*, such as personal insults or hateful slurs; *physical harassment*, such as simulated touching or grabbing; and *environmental harassment*, such as displaying graphic content on a shared screen.
>
> *(p. 2)*

They conclude that a "lack of a shared vocabulary or regulatory structure across individual social VR applications, when coupled with the novelty of the technology and the transience of virtual spaces, presents unique challenges for the development of consistent pro-social norms" (p. 24).

The Safety and Security theme includes principles that "express requirements that AI systems be safe, performing as intended, and also secure, resistant to being compromised by unauthorized parties" (Fjeld et al. 2020, p. 4). "The principle of 'safety' requires that an AI system be reliable and that 'the system will do what it is supposed to do without harming living beings or [its] environment'" (p. 38). Questions arise for contexts more specific to augmentation. What about the impact when technology is embedded in bodies, when removing or disabling devices is not possible, or when it might even be harmful to remove them (e.g., Second Sight)? Like AI systems, augmentation technologies sit in a paradoxical position in terms of safety as a concept. Many are developed in the name of safety or security by providing an embodied physical enhancement, and yet they might place human subjects in far more dangerous scenarios through their usage. Mohamed Khamis and Florian Alt discuss some of the potentially dangerous contexts:

> Augmented users will carry cameras and sensors that may allow them to spy on others (e.g., listening to distant conversations ...), infer sensitive information about them without their consent (e.g., revealing their PINs using thermal imaging (Abdelrahman et al., 2017), or even harm others (e.g., robotic limbs (Al-Sada et al., 2019) could unintentionally hit bystanders).
>
> *(Khamis & Alt, 2021, p. 271)*

Enhancement Vulnerabilities

The urgency to evaluate the safety and security of augmentation technologies is prominent in defense discourse domains, representing people and organizations that have long designed them. The UK Ministry of Defence in collaboration with Germany's office for defense planning released *Human Augmentation—The Dawn of a New Paradigm* in 2021, a long report that stated, by way of explaining its intent, that the:

> prospect of using technology to radically enhance human performance has been the subject of intense debate over the last two decades.... What is certain is that the field of human augmentation has the potential to transform society, security and defence over the next 30 years,

and further, "We must begin to understand the implications of these changes and shape them to our advantage now, before they are thrust upon us" (p. 3).

The report notes that "significant thought has already been given to what this means for artificial intelligence, automation and robotics, but comparatively little time has been given to what this means from a human perspective" (p. 3). It goes on to highlight the safety risks:

> The safety of a technology, especially one that is highly invasive and/or poorly understood, is a critical ethical constraint... augmentation technologies that entail high risks to deliver minor benefits are unlikely to be judged as worthwhile. Safety risks are particularly difficult to weigh up in a military context; for example, brain interfaces and exoskeletons designed to enhance performance (and survivability) could inadvertently create vulnerabilities to hacking, jamming and mobility in confined positions.
>
> *(p. 47)*

Safety and Security Principles

- **Safety** is defined as an AI system being "reliable and that 'the system will do what it is supposed to do without harming living beings or [its] environment'" (Fjeld et al., 2020, p. 38).
- **Security** "concerns an AI system's ability to resist external threats" and "for three specific needs to protect against security threats: the need to test the resilience of AI systems; to share information on vulnerabilities and cyberattacks; and to protect privacy and 'the integrity and confidentiality of personal data'" (Fjeld et al., 2020, p. 39).
- **Security by Design** is "related to the development of secure AI systems... [it] may provide a link between abstract principles and specific implementation decisions" (Fjeld et al., 2020, pp. 39–40). It suggests that security principles should actually be implemented.
- **Predictability** means that "for a system to be predictable, the outcome of the planning process must be consistent with the input... with the Beijing AI Principles observing that improving predictability, alongside other 'ethical design approaches' should help 'to make the system trustworthy'" (Fjeld et al., 2020, p. 40).

Theme 4: Transparency and Explainability

Pew Research Center published a report called *Artificial Intelligence and the Future of Humans* that surveyed "979 technology pioneers, innovators, developers, business and policy leaders, researchers and activists." The report opens, "Digital life is augmenting human capacities and disrupting eons-old human activities... most experts, regardless of whether they are optimistic or not, expressed concerns about the long-term impact of these new tools on the essential elements of being human" (Anderson et al., 2018, p. 2). What are the consequences when information about augmentation technologies is not made transparent by those entities that develop them, such as businesses, universities, or governments? The most fundamental consequence is that none of the other principles can be implemented. Safety and security oversight cannot be ensured if augmentation systems are not made transparent for users. Or equally important, without explainable, transparent systems and populations that understand them, systems that benefit people will never be deployed. danah boyd, principal researcher for Microsoft and founder and president of the Data & Society Research Institute, states:

> AI is a tool that will be used by humans for all sorts of purposes, including in the pursuit of power. There will be abuses of power that involve AI, just as there will be advances in science and humanitarian efforts that also involve AI. Unfortunately, there are certain trend lines that are likely to create massive instability.
>
> *(Anderson et al., 2018, p. 6)*

Like AI technologies, most augmentation technologies are in the pre- or early release stages. Throughout this book, we are dedicated to positioning TPCs to understand augmentation technologies to design ethical futures. A key component for understanding these technologies is making the implications of adopting them fully transparent, so that the full range of consequences can be addressed early in the design cycle. We dedicate space for this theme in Chapter 4, which discusses explainability and the importance of requiring designers, developers or the technologies themselves, to explain how they work. To fully support transparency and explainability, we explore and promote explainable AI (XAI), digital literacy, AI literacy, and other competencies in detail through the lens of several experts. We note the novel difficulty that AI poses for transparency and explainability:

> The lack of understanding about how an AI system works, in some cases even by the people who have developed it, is one of the reasons AI poses novel safety, ethical, and legal considerations, and why oversight and governance are especially important.
>
> *(Newman, 2021)*

Transparency and Explainability Principles

AI Principles under the Transparency and Explainability theme "articulate requirements that AI systems be designed and implemented to allow for oversight, including through translation of their operations into intelligible outputs and the provision of information about where, when, and how they are being used" (Fjeld et al., 2020, p. 5).

- **Transparency** means "that AI systems should be designed and implemented in such a way that oversight of their operations are possible" (Fjeld et al., 2020, p. 42).
- **Explainability** is "the translation of technical concepts and decision outputs into intelligible, comprehensible formats suitable for evaluation" (Fjeld et al., 2020, p. 42).
- **Open Source Data and Algorithms** emphasizes "the value of the development of common algorithms and open research and collaboration to support the advancement of the technology" to avoid monopolies and make them open for audits and other measures (Fjeld et al., 2020, pp. 43–44).
- **Open Government Procurement** is:

  the requirement that governments be transparent about their use of AI systems…. The Access Now report recommends that: 'When a government body seeks to acquire an AI system or components thereof, procurement should be done openly and transparently according to open procurement standards.'

  (Fjeld et al., 2020, p. 44)

- **Right to Information** is "the entitlement of individuals to know about various aspects of the use of, and their interaction with, AI systems" (Fjeld et al., 2020, 44).
- **Notification When AI Makes a Decision about an Individual** stipulates that "where an AI has been employed [to make a decision about a person], the person [involved]… should know…. If people don't know when they are subject to automated decisions, they won't have the autonomy to decide whether or not they consent, or the information to reach their own conclusions about the overall value that AI provides" (Fjeld et al., 2020, 45).
- **Notification When Interacting with AI** is the requirement "that humans should always be made aware when they are engaging with technology rather than directly with another person," and "this principle is a precondition to the actualization of other principles" such as accountability or privacy. (Fjeld et al., 2020, 46).
- **Regular Reporting** requires that "organizations that implement AI systems should systematically disclose important information about their use" (Fjeld et al., 2020, p. 46).

Theme 5: Fairness and Non-discrimination

Augmentation technologies are developed, adopted, and adapted in post-internet society that has already demonstrated problems of platforms diminishing human experiences. Augmentation technology emerges globally through rightsholders and organizations that contribute to its rise. With global platforms operating through international technical standards, industries develop software and hardware, share information, deploy methods of payments, and grow, largely without limitation by national systems of governance. One result is that citizens' rights have been significantly compromised and individuals have been harmed. From the Harvard Kennedy School, Carr Center for Human Rights Policy, Sabelo Mhlambi (2020) writes:

> The construction of value in the digital technology industry based on the commodification of users amounts to processes of reduction that lead to the isolation and dehumanization of individuals. The commodification resulting from designing one-sided objectives—objectives ultimately designed to increase a company's profits, sometimes through the capture of a user's attention—results in a diminished digital representation of ourselves and treats people as a means rather than an end. The extraction of our data reduces a holistic view of a person and leads to models designed to maximize profit.
>
> *(p. 22)*

We draw on Sabelo Mhlambi's key point that rather than enhancing life, digital technology industries can sometimes dehumanize individuals due to companies' primary motive, which is to increase profits. The value of efficiency for Mhlambi is the problematic "commodification of users" (p. 22).

Racial and Gender Discrimination

In this theme, we draw heavily on previous work conducted by the AI Now Institute at New York University because it produces "interdisciplinary research and public engagement to help ensure that AI systems are accountable to the communities and contexts in which they're applied" (ainowinstitute. org). The *Discriminating Systems: Gender, Race and Power in AI* report (West et al. 2019) explains the urgent situation:

> From a high-level view, AI systems function as systems of discrimination: they are classification technologies that differentiate, rank, and categorize. But discrimination is not evenly distributed. A steady stream of examples in recent years have demonstrated a persistent problem of gender and race-based discrimination (among other attributes and forms of identity). Image recognition technologies miscategorize black faces, sentencing algorithms discriminate against black defendants, chatbots easily adopt racist and misogynistic language when trained on online discourse, and Uber's facial recognition doesn't work for trans drivers. In most cases, such bias mirrors and replicates existing structures of inequality in society.
>
> *(p. 6)*

We pull from this paragraph "AI systems function as systems of discrimination" (p. 6) to make salient the fact that augmentation technology—informed by AI in material forms such as algorithmic bias—can also function to discriminate. The challenge is to ensure that augmentation technology is designed with ethically aligned priorities across its phases of emergence: design, adoption, and adaptation.

We also excerpt the complete list of research findings from the report (West et al., 2019) because it highlights the broad consequences associated with discrimination in the tech industry with focus on AI:

> **There is a diversity crisis in the AI sector across gender and race.** Recent studies found only 18% of authors at leading AI conferences are women, and more than 80% of AI professors are men. This disparity is extreme in the AI industry: women comprise only 15% of AI research staff at Facebook and 10% at Google. There is no public data on trans workers or other gender minorities. For black workers, the picture is even worse. For example, only 2.5% of Google's workforce is black, while Facebook and Microsoft are each at 4%. Given decades of concern and investment to redress this imbalance, the current state of the field is alarming.
>
> **The AI sector needs a profound shift in how it addresses the current diversity crisis.** The AI industry needs to acknowledge the gravity of its diversity problem, and admit that existing methods have failed to contend with the uneven distribution of power, and the means by which AI can reinforce such inequality. Further, many researchers have shown that bias in AI systems reflects historical patterns of discrimination. These are two manifestations of the same problem, and they must be addressed together.
>
> **The overwhelming focus on 'women in tech' is too narrow and likely to privilege white women over others.** We need to acknowledge how the intersections of race, gender, and other identities and attributes shape people's experiences with AI. The vast majority of AI studies assume gender is binary, and commonly assign people as 'male' or 'female' based on physical appearance and stereotypical assumptions, erasing all other forms of gender identity.
>
> **Fixing the 'pipeline' won't fix AI's diversity problems.** Despite many decades of 'pipeline studies' that assess the flow of diverse job candidates from school to industry, there has been no substantial progress in diversity in the AI industry. The focus on the pipeline has not addressed deeper issues with workplace cultures, power asymmetries, harassment, exclusionary hiring practices, unfair compensation, and tokenization that are causing people to leave or avoid working in the AI sector altogether.

The use of AI systems for the classification, detection, and prediction of race and gender is in urgent need of re-evaluation. The histories of 'race science' are a grim reminder that race and gender classification based on appearance is scientifically flawed and easily abused. Systems that use physical appearance as a proxy for character or interior states are deeply suspect, including AI tools that claim to detect sexuality from headshots, predict 'criminality' based on facial features, or assess worker competence via 'micro-expressions.' Such systems are replicating patterns of racial and gender bias in ways that can deepen and justify historical inequality. The commercial deployment of these tools is cause for deep concern.

(West et al., 2019, p. 3, used under CC BY-ND 4.0)

We highlight one point above: "The histories of 'race science' are a grim reminder that race and gender classification based on appearance is scientifically flawed and easily abused." We turn to the work of Alex Hanna, Emily Denton, Andrew Smart, and Jamila Smith-Loud, who "argue that algorithmic fairness researchers need to take into account the multidimensionality of race, take seriously the processes of conceptualizing and operationalizing race, focus on social processes which produce racial inequality, and consider perspectives of those most affected by sociotechnical systems" (Hanna et al., 2020, p. 1).

The Montreal AI Ethics Institute is another international non-profit organization that is "democratizing AI ethics literacy" (montrealethics.ai). One institute report catalogs extensive work that has been contributed by AI ethics scholars. Katlyn Turner, Danielle Wood, and Catherine D'Ignazio explain some of the issues:

In the past decade, Black women have been producing leading scholarship that challenges the dominant narratives of the AI and Tech industry: namely that technology is ahistorical, "evolved", "neutral" and "rational" beyond the human quibbles of issues like gender, class, and race. Safiya Noble demonstrates how search algorithms routinely work to dehumanize Black women and girls (Noble, 2019). Ruha Benjamin challenges what she calls the "imagined objectivity" of software and explains how Big Tech has collaborated with unjust systems to produce "the New Jim Code", software products that work to reproduce racial inequality (Benjamin, 2019). Joy Buolamwini and Timnit Gebru definitively expose racial and gender bias in facial analysis libraries and training datasets (Buolamwini & Gebru, 2018). Meredith Broussard challenges the "technochauvinism" embedded in AI and machine learning products (Broussard, 2019). Rediet Abebe calls for us to confront the limitations of the concept of fairness and center our analysis on power (Kasy & Abebe, 2020). Simone Browne teaches us that today's cutting-edge technologies are part of a long history of surveillance of Black bodies in public spaces (Browne, 2015).

(Turner et al., 2021, p. 15).

The ongoing work of the writers mentioned above contributes to the field of AI literacy and specifically AI ethics literacy. We list some of their scholarly works below to both encourage people to use them and to help decolonize AI and prevent unethical outcomes for augmentation technology:

- Joy Buolamwini and Timnit Gebru, "Gender shades: Intersectional accuracy disparities in commercial gender classification," *Proceedings of 1st Conference on Fairness, Accountability, and Transparency,* 2018.
- Ruha Benjamin, *Race After Technology: Abolitionist Tools for the New Jim Code,* 2019.
- Safiya Umoja Noble, *Algorithms of Oppression: How Search Engines Reinforce Racism,* 2019.
- Maximilian Kasy and Rediet Abebe, "Fairness, equality, and power in algorithmic decision making," *Conference on Fairness, Accountability, and Transparency,* 2021.
- Simone Browne, *Dark Matters: On the Surveillance of Blackness,* 2015.
- Meredith Broussard, *Artificial Unintelligence: How Computers Misunderstand the World,* 2019.

Ableism

In Chapter 2, we discussed issues surrounding assistive technologies, transhumanism, and ableism. Recently, the AI Now Institute published the *Disability, Bias, and AI* report and determined the following:

> Disability has been largely omitted from the AI-bias conversation, even as disabled people are affected by these issues in differing ways across axes of identity. Disabled people have been subject to historical and present-day marginalization, much of which has systematically and structurally excluded them from access to power, resources, and opportunity.
>
> *(Whittaker et al., 2019, p. 8)*

Ableism can be defined as "Discrimination in favor of the non-disabled or against people with disabilities" (Whittaker et al, 2019, p. 5.). One key AI principle is using "representative and high-quality data" to help ensure that results are not biased or causing a "skewed representation of a group" (Fjeld et al., 2020, p. 49). When dealing with augmentation technologies, defined as enhancing human capabilities, there are many sites where ableism might cause profound harm. The *Disability, Bias, and AI* report explains, "When considering

AI and disability, It's important to acknowledge this logic as core to how AI systems function. Despite the fact that there are over a billion disabled people worldwide, most social structures and institutions treat disabled people as deficient or dependent, as 'atypical' or 'abnormal'" (Whittaker et al., 2019, p. 12).

We encourage TPC scholars and professionals to draw on resources from the ACM special interest group on accessible computing (SIGACCESS):

> ACM SIGACCESS supports the international community of researchers and professionals applying computing and information technologies to empower individuals with disabilities and older adults. The SIG also promotes the professional interests of students and computing personnel with disabilities and strives to educate the public to support careers for people with disabilities.
>
> *(SIGACCESS, 2022)*

For TPC professionals, researchers, and students, SIGACCESS has published a guide to creating accessible virtual conferences inclusive for those with disabilities. Chapter 8 discusses further ways to assimilate resources into work.

Digital Divide

Will augmentation technology adoption contribute to societal inequality? Manuel Castells wrote *The Rise of the Network Society* in 1996. He is one of the first scholars to identify how emergent networking technologies helped to restructure capitalism and a global economy, resulting in what we now call "network society" (Castells, 1996, 2000). Network society caused a new global polarity between people who are inside of networked society, and those left marginalized outside of it. The term "digital divide" has been defined as "a division between people who have access and use of digital media and those who do not" (van Dijk, 2017). But the consequences are bound to states of inequality:

> The digital divide refers to the inequality of access of people and places to digital technologies, particularly the Internet. There are different types of divides: social, based on factors like income, gender, [race,] and age; or spatial such as distinctions in access between urban and rural communities, regions of a country, or parts of the world.
>
> *(Castree et al., 2013)*

Critical media studies have proven how large tech companies have advanced global communication infrastructures over the past 20 years to empower corporations and exploit individuals (Couldry & Mejías, 2019). In Chapter 3, we discuss how the Metaverse concept leverages Big Tech corporate ambitions in post-internet society and how it will contextualize augmentation technology in the future. We argue that knowing about digital divides is tied to competencies we discuss in Chapter 4, the ability to critically evaluate AI technologies for their ethical consequences.

Journalist Karen Hao has written an important set of articles called *MIT Technology Review's series on AI colonialism*; the first article, "Artificial intelligence is creating a new colonial world order" (Hao, 2022), discusses the series:

> Together, the stories reveal how AI is impoverishing the communities and countries that don't have a say in its development—the same communities and countries already impoverished by former colonial empires. They also suggest how AI could be so much more—a way for the historically dispossessed to reassert their culture, their voice, and their right to determine their own future.

Through the voices of these community members in the global south in the articles, we learn more about decolonizing AI and the need for ethically aligned design to overcome divides. Engineers and ethicists are using augmentation technologies from inside their communities and in collaboration with others to bring about change, and for example, to revive indigenous culture responsibly and ethically (Holderman et al., 2022).

Governments around the world are acknowledging the fact that access to AI techniques, technologies and applications could contribute to further inequities and digital divides. Japan's principles make specific reference to digital divides:

> The Japanese AI principles, while acknowledging the economic dimension of this issue (observing that 'AI should not generate a situation where wealth and social influence are unfairly biased towards certain stakeholders'), emphasize the sociopolitical dimensions of inequality, including the potential that AI may unfairly benefit certain states or regions as well as contribute to 'a digital divide with so-called 'information poor' or 'technology poor' people left behind.'
>
> *(Fjeld et al., 2020, p. 62)*

The notions of "information poor" or "technology poor" and other forms of sociopolitical dimensions of inequality caused by AI technologies are documented in many of the national and international multi-stakeholder principles documents.

Enhancement technologies have long been discussed in terms of societal inequality and future class divides. Patrick Lin and his colleagues discuss the ethical, legal, and operational implications of so-called super soldiers and make the point that these technologies will impact society at large:

There may be friction between the enhanced and unenhanced, or at least a class divide—in terms of education, job outlook, etc.—as we already see between those with Internet access and those without. If enhancements in society are expensive and only afforded by the wealthier, then this may widen the gap between the haves and the have-nots. Similarly, would there be a communication divide between the enhanced and unenhanced.

(Lin et al., 2014, p. 154)

In Chapter 1, we discuss the connections between the recent drive for augmentation technology and the longer history of technologies associated with "superhuman" innovation. Augmentation technologies (cognitive, sensory, physical, and emotional enhancing) and related communication practices are implicated in the widening gap between the haves and the have-nots. Likewise, the Canadian government states that "if access to health-improving or ability-enhancing augmentations is reserved only for wealthy buyers, the gap between rich and poor could further expand by enabling the wealthy to live longer and perform better than those unable to afford enhancements" (Policy Horizons Canada, 2018), a situation which would not meet fairness expectations for Canadian citizens.

Fairness and Non-discrimination Principles

- **Non-discrimination and the Prevention of Bias** holds "that bias in AI—in the training data, technical design choices, or the technology's deployment—should be mitigated to prevent discriminatory impacts" (Fjeld et al., 2020, p. 48).
- **Algorithmic Bias** is "the systemic under- or over-prediction of probabilities for a specific population" (Fjeld et al., 2020, p. 47).
- **Representative and High-Quality Data** calls for the "use of appropriate inputs to an AI system, which relates accurately to the population of interest. The use of a dataset that is not representative leads to skewed representation of a group in the dataset compared to the actual composition of the target population, introduces bias, and reduces the accuracy of the system's eventual decisions" (Fjeld et al., 2020, p. 49).
- **Fairness** is "defined as equitable and impartial treatment of data subjects by AI systems" (Fjeld et al., 2020, p. 49).
- **Equality** "stands for the idea that people, whether similarly situated or not, deserve the same opportunities and protections with the rise of AI technologies" (Fjeld et al., 2020, p. 50).
- **Inclusiveness in Impact** "calls for a just distribution of AI's benefits, particularly to populations that have historically been excluded" (Fjeld et al., 2020, p. 51).

- **Inclusiveness in Design** is "the idea that ethical and rights-respecting AI requires more diverse participation in the development process for AI systems" (Fjeld et al., 2020, p. 51).

Theme 6: Human Control of Technology

One core ethical debate about *augmenting* human skills and capacities is based on the moral and existential question of relinquishing control to a machine. *Giving up control* might involve myriad activities. It may involve some form of human-AI teaming, whereby a non-human agent takes on a role in a professional context, or it might mean allowing a sensing technology to read, interpret, and decide an outcome for a human participant using a heightened sensory ability. Like so many of the principles we discuss in this chapter, human control of technology is problematic. In AI systems development, "clear requirements for researchers, designers, and engineers are yet inexistent, making the development of AI-based systems that remain under meaningful human control challenging" (n.p., Siebert et al., 2022).

Human control of technology sits at the center of the augmentation technology paradigm itself. The machine-autonomy versus human control debate, with its overlap in moral machines research, is discussed often in reference to self-driving vehicles, autonomous weapons, and robots. Fueled by both popular science hype and substantive scholarship, these fields lead the public conversation over human control because they can be so controversial. Yet, agency and control go hand-in-hand, as we have explored in Chapters 3 and 4 and will address further in Chapter 7.

Human Control of Technology Principles

- **Human Review of Automated Decision** is "the idea that where AI systems are implemented, people who are subject to their decisions should be able to request and receive human review of those decisions" (Fjeld et al., 2020, p. 53).
- **Ability to Opt Out of Automated Decisions** is "affording individuals the opportunity and choice not to be subject to AI systems where they are implemented" (Fjeld et al., 2020, p. 54).
- **Human Control of Technology** "requires that AI systems are designed and implemented with the capacity for people to intervene in their actions." (Fjeld et al., 2020, p. 54).

Theme 7: Professional Responsibility

Professional responsibility recognizes "the vital role that individuals involved in the development and deployment of AI systems play in the systems' impacts, and call on their professionalism and integrity in ensuring that the appropriate stakeholders are consulted and long-term effects are planned for" (Fjeld et al., 2020, p. 5). We discuss *responsible AI* often in this book and promote the need to adopt its premises.

While this theme appears to be unambiguous, it is one of the most controversial of them all when contextualized in regards to augmentation technologies. In their article "The Internet of Bodies: Life and Death in the Age of AI" in *California Western Law Review,* Elinore Pauwels and Sarah Denton write:

> The fourth industrial revolution—characterized by technological advances in genomics and cyber-technologies including artificial intelligence ("AI"), automation, gene-editing, and neuroscience—seeks to merge our physical, digital, and biological lives. Yet, the Internet of Bodies exposes us to unprecedented privacy and cybersecurity vulnerabilities, introducing conflict across regulatory regimes. Now, societies must open a dialog to begin identifying the human values and norms that will define responsible AI governance and data optimization.
>
> *(Pauwels & Denton, 2018, p. 221)*

The Internet of Bodies (IoB) is a subfield of the Internet of Things (IoT) used for the ambient surveillance of bodies across wireless connections. To consider the long-term effects of augmentation technology, we will have to reenvision the way we assess our adoption practices.

Facial recognition systems have been proven to misidentify people of color, causing discriminatory outcomes (Harwell, 2019). In Chapter 1, we use facial recognition software company Clearview AI as an example of how a business model operates outside the scope of privacy regulations and conducts irresponsible professional practices. The Clearview AI website describes the company as "providing the most cutting-edge technology to law enforcement to investigate crimes, enhance public safety and provide justice to victims" (Clearview AI, 2022). Clearview AI scraped billions of facial images from social media platforms and used them in products sold to clients; its use is "illegal in Canada, Australia and parts of Europe for violating privacy laws" (Mac & Hill, 2022). In 2022, Clearview AI "settled a lawsuit brought by the American Civil Liberties Union and agreed to limit its face database in the United States primarily to government agencies and not allow most American companies to have access to it" (Mac & Hill, 2022). This example demonstrates professional responsibility as an ethical theme; there are ethical violations in specific principles including technological accuracy, scientific integrity, and ignoring the potential for long-term harms, leading to racial discrimination.

Linked to both the themes of Professional Responsibility (Consideration of Long-Term Effects principle) and Human Control of Technology (Human Review of Automated Decision principle) is the problem with surveillance platforms. State and corporate actors are involved in the formation of large surveillance platforms for preemptive decision-making. Using quantitative analysis of patents with topic modeling, Andrew Iliadis and Amelia Acker illustrate how biometric data firm Palantir "wants its surveillance platform to become the de facto operating system for governments and the world's most sensitive data across private industries and domains while also ingesting open data from public institutions" (Iliadis & Acker, 2022, n.p.).

Professional Responsibility Principles

- **Accuracy** means that an AI technology should "correctly classify information into the correct categories, or [exhibit] its ability to make correct predictions, recommendations, or decisions based on data or models" (Fjeld et al., 2020, p. 57).
- **Responsible Design** is the notion that "individuals must be conscientious and thoughtful when engaged in the design of AI systems... there is a strong consensus that professionals are in a unique position to exert influence on the future of AI" (Fjeld et al., 2020, p. 57).
- **Consideration of Long-Term Effects** concerns the "deliberate attention to the likely impacts, particularly distant future impacts, of an AI technology during the design and implementation process... AI capabilities in the future may be vastly advanced compared to the technology we know today" (Fjeld et al., 2020, p. 59).
- **Multi-stakeholder Collaboration** involves "encouraging or requiring that designers and users of AI systems consult relevant stakeholder groups while developing and managing the use of AI applications" (Fjeld et al., 2020, p. 59).
- **Scientific Integrity** advocates that "those who build and implement AI systems should be guided by established professional values and practices" (Fjeld et al., 2020, p. 59).

Theme 8: Promotion of Human Values

How will communicators work in contexts that are morally challenging when values about augmentation technologies are not shared or agreed upon? The *Principled Artificial Intelligence* report expresses that "the ends to which AI is devoted, and the means by which it is implemented, should correspond with our core values and generally promote humanity's well-being" (Fjeld et al., 2020, p. 5). Virginia Dignum, in her book *Responsible Artificial Intelligence* (2019), writes "Determining what is 'good' means that one needs to know what are the underlying values. Each person and socio-cultural environment prioritises and interprets moral and societal values differently" (p. 35).

The way cultures codify, communicate, and police adherence to human value systems is multidimensional and one can never assume it to be homogenous. In 2017, Nadja Oertelt and her colleagues formed a group and proposed *Human by Design: An Ethical Framework for Human Augmentation* in IEEE *Technology and Society*. They suggest "navigating ethical questions that emerge around human augmentation" for stakeholder communities (Oertelt et al., 2017). Geared to militaries and designing for a value-based paradigm, they provide questions for military personnel who could be made vulnerable during the adoption of these technologies within their professional contexts:

- Can the military ask—or order—soldiers to augment their bodies in ways that are irreversible?
- Can the military ask, or order, soldiers to augment themselves in ways that make them more likely to act in certain ways?
- Is it immoral to augment the capacities of soldiers to make them more effective or efficient during times of war?
- How can the military accommodate or facilitate the reintegration of an augmented soldier into civilian population if they have been augmented with conspicuously different capacities? (Oertelt et al., 2017).

While codes of ethics for military personnel already circumscribe their moral and professional behavior (e.g., Department of National Defence and Canadian Forces Code of Values and Ethics), augmentation technologies push moral boundaries. Will augmentation technology lead to the instrumentalization of enhanced people? Narratives implied in these questions often appear in sci-fi movies exploring technological disruption associated with military–industrial complexes and value systems.

TPC professionals work on global teams in highly globalized contexts. We raised the socio-ethical consequence of "digital divides" above and we revisit it here because the principle of "access to technology" is a human value and increasingly considered a human right. Access to technology may involve training, education, and technical communication.

Promotion of Human Values Principles

- **Human Values and Human Flourishing** is "the development and use of AI with reference to prevailing social norms, core cultural beliefs, and humanity's best interests" (Fjeld et al., 2020, p. 61).
- **Access to Technology** means the "broad availability of AI technology, and the benefits thereof, is a vital element of ethical and rights-respecting AI" (Fjeld et al., 2020, p. 61).
- **Leveraged to Benefit Society** means that "AI systems should be employed in service of public-spirited goals" (Fjeld et al., 2020, p. 62).

International Human Rights

In November 2021, the United Nations organized 193 countries to adopt the first global Recommendation on the Ethics of Artificial Intelligence. The first clause in its preamble is:

> Recognizing the profound and dynamic impact of artificial intelligence (AI) on societies, ecosystems, and human lives, including the human mind, in part because of the new ways in which it influences human thinking, interaction and decision-making and affects education, human, social and natural sciences, culture, and communication and information...
>
> *(UNESCO, 2021)*

This chapter has explained the broad sphere of socio-ethical consequences surrounding augmentation technologies and the need to be knowledgeable of social harms now and in the future. There are personal and professional affordances and positive social consequences that we are also promoting. For instance, Dane Acena and Guo Freeman (2021) argue for the unique affordances of social VR, stating that "despite limited LGBTQ participants, prior research has found that social VR provides transgender (trans) users with an immersive and embodied way to explore, express, and experiment with their gender" (p. 1). They discuss designing safe spaces for transgender children to interact with other children in virtual worlds. Virtual worlds and ambient augmentation technologies offer "a more immersive 360-degree space for interaction and allow them to interact via voice, embodied avatars through full-body tracking (i.e., avatar's movements correspond to one's physical body movements), and more customized self-presentation through their virtual avatar" supporting "LGBTQ users' social and identity practices beyond what conventional social media platforms or traditional online gaming/virtual worlds could afford" (Acena & Freeman, 2021, p. 2).

Chapter 6 integrates the study of augmentation technologies for technical communication undergraduate and graduate curricula, with a discussion of learner capabilities. Chapter 7 provides background on emerging professional practices for the numerous fields adopting and adapting to augmentation technologies.

References

Abdelrahman, Y., Khamis, M., Schneegass, S., & Alt, F. (2017). Stay cool! Understanding thermal attacks on mobile-based user authentication. In *Proceedings of the 35th annual ACM conference on human factors in computing systems, CHI '17*. ACM, New York, NY, USA. https://doi.org/10.1145/3025453.3025461

Acena, D., & Freeman, G. (2021). 'In my safe space': Social support for LGBTQ users in social virtual reality. In *Extended abstracts of the 2021 CHI Conference on*

Human Factors in Computing Systems (CHI EA '21), Article 301, 1–6. https://doi. org/10.1145/3411763.3451673

Al-Sada, M., Höglund, T., Khamis, M., Urbani, J., & Nakajima, T. (2019). Orochi: Investigating requirements and expectations for multipurpose daily used supernumerary robotic limbs. In *Proceedings of the 10th augmented human international conference 2019, AH 2019*. ACM, New York, NY, USA, pp. 37:1–37:9. https://doi. org/10.1145/3311823.3311850

Anderson, J., Rainie, L., & Luchsinger, A. (2018). *Artificial intelligence and the future of humans*. Pew Research Center.

Andrejevic, M. (2007). *iSpy: Surveillance and power in the interactive era*. University Press of Kansas.

Armis. (2017). *BlueBorne*. https://www.armis.com/blueborne/#/general

Benjamin, R. (2019). *Race after technology: Abolitionist tools for the new Jim Code*. Polity.

Blackwell, L., Ellison, N., Elliott-Deflo, N., & Schwartz, R. (2019). Harassment in social virtual reality: Challenges for platform governance. In *Proceedings of the ACM on human-computer interaction, 3*(CSCW), 1–25.

British Broadcasting Corporation (2022, May 25). Female avatar sexually assaulted in Meta VR platform, campaigners say. *BBC News*. https://www.bbc.com/news/technology-61573661

Broussard, M. (2019). *Artificial unintelligence: How computers misunderstand the world*. MIT Press.

Browne, S. (2015). *Dark matters: On the surveillance of blackness*. Duke University Press.

Buolamwini, J., & Gebru, T. (2018). Gender shades: Intersectional accuracy disparities in commercial gender classification. In *Proceedings of 1st conference on fairness, accountability, and transparency* (pp. 77–91).

Cameron, K. (2021, November 15). *Is it OK to leave Bluetooth on all of the time?* The Gadget Buyer. https://thegadgetbuyer.com/is-it-ok-to-leave-bluetooth-on-all-of-the-time/

Castells, M. (1996). *The rise of the network society*. Blackwell.

Castells, M. (2000). Toward a sociology of the network society. *Contemporary Sociology, 29*(5), 693–699.

Castree, N., Kitchin, R., & Rogers, A. (2013) Digital divide. *Oxford dictionary of human geography*. Oxford University Press.

Clearview AI (2022). https://www.clearview.ai/

Couldry, N., & Mejías, U. A. (2019). *The costs of connection: How data is colonizing human life and appropriating it for capitalism*. Stanford University Press.

Crampton, J. W. (2019). Platform biometrics. *Surveillance & Society, 17*(1/2), 54–62.

Crawford, K. (2021). *Atlas of AI: Power, politics and the planetary costs of artificial intelligence*. Yale University Press.

Czerwinski, M., Hernandez, J., & McDuff, D. (2021). Building an AI that feels. *IEEE Spectrum, 58*(5), 32–38.

Dignum, V. (2019). *Responsible artificial intelligence: How to develop and use AI in a responsible way*. Springer. https://doi.org/10.1007/978-3-030-30371-6_3

Dingler, T., Niforatos, E., & Schmidt, A. (2021). From toolmakers to cyborgs. In T. Dingler & E. Niforatos (Eds.), *Technology-augmented perception and cognition* (pp. 1–10). Springer Nature.

Duin, A. H., & Pedersen, I. (2023). *Socio-ethical consequences and design futures* [Multimedia collection]. Fabric of Digital Life. https://fabricofdigitallife.com/Browse/objects/facets/collection:73

Fjeld, J., Achten, N., Hilligoss, H., Nagy, A., & Srikumar, M. (2020). *Principled artificial intelligence: Mapping consensus in ethical and rights-based approaches to principles for AI.* Berkman Klein Center Research. https://ssrn.com/abstract=3518482

Frenkel, S., & Browning, K. (2021, December 30). The metaverse's dark side: Here come harassment and assaults. *New York Times.* https://www.nytimes.com/2021/12/30/technology/metaverse-harassment-assaults.html

Grover, T., Rowan, K., Suh, J., McDuff, D., & Czerwinski, M. (2020). Design and evaluation of intelligent agent prototypes for assistance with focus and productivity at work. In *Proceedings of the 25th international conference on Intelligent User Interfaces* (IUI '20) (pp. 390–400).

Hanna, A., Denton, E., Smart, A., & Smith-Loud, J. (2020). *Towards a critical race methodology in algorithmic fairness.* Conference on Fairness, Accountability, and Transparency (FAT★ '20), January 27–30, 2020, Barcelona, Spain. https://doi.org/10.1145/3351095.3372826

Hao, K. (2022, April 19). Artificial intelligence is creating a new colonial world order. *MIT Technology Review.* https://www.technologyreview.com/2022/04/19/1049592/artificial-intelligence-colonialism/

Hartzog, W. (2020). BIPA: The most important biometric privacy law in the US? In A. Kak (Ed.), *Regulating biometrics: Global approaches and urgent questions* (pp. 96–103). AI NOW Institute. https://ainowinstitute.org/regulatingbiometrics.html

Harwell, D. (2019, December 21). Federal study confirms racial bias of many facial-recognition systems, casts doubt on their expanding use. *Washington Post.* https://www.washingtonpost.com/technology/2019/12/19/federal-study-confirms-racial-bias-many-facial-recognition-systems-casts-doubt-their-expanding-use/

Holderman, C., Jain, E., Running Wolf, M., & Erickson, L. (2022). Privacy, safety and wellbeing: Solutions for the future of AR and VR. In *Proceedings of the 35th annual conference of the Special Interest Group on Computer Graphics and Interactive Techniques* (SIGGRAPH '22). ACM, New York, NY, USA, Article 1, 1–2. https://doi.org/10.1145/3532718.3535620

Iliadis, A. (2020). Computer guts and swallowed sensors: Ingestibles made palatable in an era of embodied computing. In I. Pedersen & A. Iliadis (Eds.), *Embodied computing: Wearables, implantables, embeddables, ingestibles* (pp. 1–20). MIT Press.

Iliadis, A., & Acker, A. (2022). The seer and the seen: Surveying Palantir's surveillance platform. *Information Society, 38*(5), 334–363.

Jain, A. K., Flynn, P., & Ross, A. A. (2010). *Handbook of biometrics.* Springer Publishing Company.

Kak, A. (Ed.). (2020). *Regulating biometrics: Global approaches and urgent questions.* AI NOW Institute. https://ainowinstitute.org/regulatingbiometrics.html

Kasy, M., & Abebe, R. (2021). Fairness, equality, and power in algorithmic decision making. In Proceedings of the 2021 *ACM conference on fairness, accountability, and transparency.* (pp. 576–586).

Khamis, M., & Alt, F., (2021). Privacy and Security in Augmentation Technologies. In T. Dingler & E. Niforatos (Eds.), *Technology-augmented perception and cognition* (pp. 71–124). Springer Nature. https://doi.org/10.1007/978-3-030-30457-7_4

Kindt, E. (2020). A first attempt at regulating biometric data in the European Union. In A. Kak (Ed.), *Regulating biometrics: Global approaches and urgent questions* (pp. 62–69). AI NOW Institute. https://ainowinstitute.org/regulatingbiometrics.html

Lin, P., Mehlman, M., Abney, K., French, S., Vallor, S., Galliott, J., Burnam-Fink, M., LaCroix, A. R., & Schuknecht, S. (2014). Super soldiers (Part 2): The ethical, legal, and operational implications. In S. J. Thompson (Ed.), *Global issues and ethical considerations in human enhancement technologies* (pp. 139–160). IGI Global. https://doi.org/10.4018/978-1-4666-6010-6.ch008

Mac, R., & Hill, K. (2022, May 9). Clearview AI settles suit and agrees to limit sales of facial recognition database. *New York Times.* https://www.nytimes.com/2022/05/09/technology/clearview-ai-suit.html

Martin, A. K., & Whitley, E. A. (2013). Fixing identity? Biometrics and the tensions of material practices. *Media, Culture & Society, 35*(1), 52–60. https://doi.org/10.1177/0163443712464558

Mhlambi, S. (2020). *From rationality to relationality: Ubuntu as an ethical and human rights framework for artificial intelligence governance.* Carr Center for Human Rights Policy. https://carrcenter.hks.harvard.edu/files/cchr/files/ccdp_2020-009_sabelo_b.pdf

Michael, K., Michael, M. G., Perakslis, C., & Abbas, R. (2020). Überveillance and the rise of last-mile implantables. In I. Pedersen & A. Iliadis (Eds.), *Embodied computing: Wearables, implantables, embeddables, ingestibles* (pp. 97–130). MIT Press.

Ministry of Defence, UK. (2021). *Human augmentation—the dawn of a new paradigm.* gov.uk. https://www.gov.uk/government/publications/human-augmentation-the-dawn-of-a-new-paradigm

Newman, J. (2021). *Explainability won't save AI.* Tech Stream. https://www.brookings.edu/techstream/explainability-wont-save-ai/

Noble, S. U. (2019). *Algorithms of oppression: How search engines reinforce racism.* NYU Press.

Oertelt, N., Arabian, A., Brugger, E. C., Farahany, N. A., Payne, S., & Rosellini, W. (2017, September 14). Human by design: An ethical framework for human augmentation. *IEEE Technology and Society.* https://technologyandsociety.org/human-by-design-an-ethical-framework-for-human-augmentation/

Parker, K., Horowitz, J. M., & Minkin, R. (2022). *COVID-19 pandemic continues to reshape work in America.* Pew Research Center. https://www.pewresearch.org/social-trends/2022/02/16/covid-19-pandemic-continues-to-reshape-work-in-america/

Pauwels, E. and Denton, S. (2018). The Internet of bodies: Life and death in the age of AI. *California Western Law Review, 55*(1), 221–233.

Pedersen, I. (2020). Will the body become a platform? Body networks, datafied bodies, and AI futures. In I. Pedersen & A. Iliadis (Eds.), *Embodied computing: Wearables, implantables, embeddables, ingestibles* (pp. 21–47). MIT Press.

Policy Horizons Canada. (2018). *The next generation of emerging global challenges: A Horizons 2030 perspective on research opportunities.* https://horizons.gc.ca/wp-content/uploads/2018/10/SSHRC-Emerging-Global-Challenges-ENG-Web-New-1.pdf

Siebert, L. C., Lupetti, M.L., & Aizenberg, E., Beckers, N., Zgonnikov, A., Veluwenkamp, H., Abbink, D., Giaccardi, E., Houben, G.-J., Jonker, C. M., van den Hoven, J., Forster, D., & Lagendijk, R. L. (2022, May 18). Meaningful human control: Actionable properties for AI system development. *AI and Ethics.* https://doi.org/10.1007/s43681-022-00167-3

SIGACCESS (2022). *Welcome to SIGACCESS.* http://www.sigaccess.org/

Strickland, E., & Harris, M. (2022, February 15). Their bionic eyes are now obsolete and unsupported. *IEEE Entrepreneurship.* https://spectrum.ieee.org/bionic-eye-obsolete

Tech at Meta (2021, March 18). Facebook Reality Labs—Human-computer interaction at the wrist [Video]. Facebook. https://www.facebook.com/316002512431444/videos/1146468922460553

Turner, K., Wood, D., & D'Ignazio, C. (2021, January). The abuse and misogynoir playbook. *The state of AI ethics report*. Montreal AI Ethics Institute (MAIEI).

UNESCO (2021, November 23). *Recommendation on the ethics of artificial intelligence.* https://unesdoc.unesco.org/ark:/48223/pf0000381137

van Dijk, J. A. G. M. (2014). Datafication, dataism and dataveillance: Big data between scientific paradigm and ideology. *Surveillance & Society, 12*(2), 197–208. https://doi.org/10.24908/ss.v12i2.4776

van Dijk, J. A. G. M. (2017). Digital divide: Impact of access. In P. Rössler, C. A. Hoffner, & L. V. Zoonen (Eds.), *The international encyclopedia of media effects*. John Wiley & Sons. https://doi.org/10.1002/9781118783764.wbieme0043

West, S.M., Whittaker, M., & Crawford, K. (2019). *Discriminating systems: Gender, race and power in AI.* AI NOW Institute.

Whittaker, M., Alper, M., Bennett, C. L., Hendren, S., Kaziunas, L., Mills, M., Morris, M. R., Rankin, J., Rogers, E., Salas, M., & West, S. M. (2019). *Disability, bias, and AI.* AI NOW Institute. https://ainowinstitute.org/disabilitybiasai-2019.pdf

SECTION THREE
Design Ethical Futures

Section Three provides direction for cultivating digital literacy and AI literacy skills needed to assess integration and participation with augmentation technologies, chronicles the rapid increase in autonomous agents and digital employees, positions TPCs for intervening throughout the design, adoption, and adaptation of augmentation technologies, and shares strategic and tactical approaches for designing ethical augmentation technology and AI futures.

DOI: 10.4324/9781003288008-8

6

PEDAGOGICAL DIRECTION FOR CULTIVATING AUGMENTATION TECHNOLOGY AND AI LITERACIES

Overview

Chapter 6 provides pedagogical direction for cultivating digital literacy and artificial intelligence (AI) literacy skills needed to assess integration and participation with augmentation technologies. It draws on multiple studies on building digital literacy (Burnham & Tham, 2021; Davis et al., 2021; Duin et al., 2021; Tham et al., 2021). It shares student and instructor perception of digital and AI literacy through studies of learner engagement with the Fabric of Digital Life cultural repository of artifacts. It discusses theoretical and pragmatic model development for instructor-scholars for student learning needs as well as their professional futures and workforce preparedness. It provides suggestions for integrating study of augmentation technologies in technical communication across undergraduate and graduate curricula. The chapter concludes with discussion of learner capabilities for a world with augmentation technologies and AI.

Key Questions

- How might instructors cultivate student digital and AI literacy skills needed to assess integration and participation with augmentation technologies?
- How have instructor-scholars used the Fabric of Digital Life (https://fabricofdigitallife.com) to cultivate digital literacy?
- What curricular direction might best promote understanding of augmentation technologies and AI across undergraduate and graduate programs in technical and professional communication?
- How might students use augmentation technologies and AI to learn independently of program curricula? What capabilities are needed for learning within a world of augmentation technologies and AI?

DOI: 10.4324/9781003288008-9

Chapter 6 Links

Throughout the chapter, we refer to articles, videos, and reports which can be found in a related chapter collection at Fabric of Digital Life called *Pedagogical direction for cultivating augmentation technology and AI literacies* (Duin & Pedersen, 2023). You can find a link to this collection at https://fabricofdigitallife.com.

Introduction

The biennial International Networked Learning conference, which first took place in 1998, brings together educational and organizational research on formal and informal technology enhanced learning settings (https://www.networkedlearning.aau.dk/). Particularly concerned with critical perspectives and empirical analysis, at a recent meeting, and together with Jason Tham, we presented "Building Digital Literacy Through Exploration and Curation of Emerging Technologies: A Networked Learning Collaborative" (Duin et al., 2021). Our opening statement presents the rationale for cultivating digital and AI literacy:

> People readily consume an ever growing range of emerging technologies while largely unaware of their lack of control over the impact that such networking, devices, data, and processes have on their lives. Massive amounts of data are collected, mined, and used to alter human behavior. In higher education and the public sphere, information about emerging technologies is often proprietary and withheld from citizens or is too complex for people to understand. As college-educated people are huge consumers of digital products which affect their own digital lives (Pedersen & Aspevig, 2018), we see it as most critical to foster student development of an expanded understanding of their digital literacy.
>
> *(p. 93)*

Amid continued study of augmentation technologies and detail on cognitive, sensory, emotional, and physical enhancements, we emphasize throughout this book the need to foster expanded understanding of digital and AI literacy.

In Chapter 4 we outlined competencies and design considerations for work with augmentation technologies, paying critical attention to definitions of digital, data and AI literacy. For example, the Museum and Library Services Act of 2010 (Pub.L. 111-340, 22 Dec. 2010) defines digital literacy as "the skills associated with using technology to enable users to find, evaluate, organize, create, and communicate information; and developing digital citizenship and the responsible use of technology" (ALA, 2022). In earlier work (Tham et al., 2021), drawing on the European Union's DigEuLit project, we noted

Allen Martin and Jan Grudziecki's (2006) definition of digital literacy as the cultivation of:

> awareness, attitude and ability... to appropriately use digital tools and facilities to identify, access, manage, integrate, evaluate, analyse and synthesize digital resources, construct new knowledge, create media expressions, and communicate with others, in the context of specific life situation[s], in order to enable constructive social action; and to reflect upon this process.
>
> *(p. 255)*

We have used this definition in our teaching and research, as it provides "a global description of digital literacy that focuses on the learner's development of knowledge, skills, and social sensitivity in dealing with digital communication" (p. 2).

Across writing studies and technical and professional communication, scholars have discussed digital literacy in terms of hypertext and cultural literacy (Tuman, 1992), information literacy (Clark, 1995), socio-technological theories (Duin & Hansen, 1996), materiality of literacy (Haas, 1996), cyberliteracy (Gurak, 2001), technological literacy (Breuch, 2002; Hovde & Renguette, 2017), layered or multiliteracies (Cargile Cook, 2002; Selber, 2004), "electracy" (Ulmer, 2003), metadata literacy (Iliadis et al., 2021), and code/coding literacy (Vee, 2017; Duin & Tham, 2018; Byrd, 2020). We emphasize that "given the ever-growing, ever-evolving nature of digital technologies, it remains daunting and confusing for students and instructors alike to understand what digital literacy encapsulates" (Tham et al., 2021, p. 2).

The field of AI has emerged amid this evolutionary definition and study of digital literacy. Barry Dainton et al. (2021) write:

> Over the course of its brief seventy-year history the field of artificial intelligence (AI) has known a succession of 'golden ages' during which advances are rapidly made, and 'ice ages' when progress has disappointingly slowed. Most commentators would agree that we are currently in the midst of an AI golden age.
>
> *(p. 2)*

Augmentation technologies and their machine intelligences do not yet possess human-level intelligence; however, cognitive, physical, emotional, and sensory enhancements continue to evolve in development of more efficient, automated systems, and emerging generative AI tools such as OpenAI's ChatGPT astound many in the ability to generate text. Dainton et al. continue:

> Humans are smart, but not that smart. It would be great to have someone a good deal smarter than us to help solve pressing problems such as

climate change, finding a cure for cancer, and reconciling quantum mechanics with general relativity—all problems which continue to defeat the most brilliant human minds. Hence there is a powerful impetus not to stop at creating AIs with human-level intelligence, but to aim for AI's that are *superintelligent*, AIs that possess vastly more intelligence than any human.

(pp. 11–12)

These researchers focus on the science fiction genre to cultivate greater understanding of AI, identifying the rapidly closing gap between science fiction and science fact: "Science fiction doesn't just have the potential to influence current thinking on AI and robotics, in many areas it has *already* exerted a very considerable influence" (p. 12). In a similar way, science fiction artifacts in the Fabric of Digital Life can be examined to cultivate digital and AI literacy for both present and imagined futures.

Use of Fabric to Build Literacies

Digital and AI literacies are prerequisites for understanding and working with this evolution of augmentation technologies. Chapter 2 illustrated how the Fabric of Digital Life archive's rich metadata fields serve as a way to examine augmentation technologies and the complex sociotechnical tradeoffs that technical and professional communicators must navigate as they work to develop usable content and direction. This chapter begins with an overview of multiple years of pedagogical research on building digital literacy (Burnham & Tham, 2021; Davis et al., 2021; Duin et al., 2021; Tham et al., 2021) to provide direction for cultivation of digital and AI literacy through examining and curating augmentation technologies.

Collections at the Fabric of Digital Life are curated by both established and student researchers. As explained previously:

Fabric's aim is to contextualize emergence within both traditional and non-traditional media genres such as magazine journalism, broadcast news, marketing outlets, tradeshow videos, video games, government publications, films, and academic research venues to reveal how digital technology is evolving. For example, an invention might be announced in an academic journal article, celebrated in a popular science magazine, and depicted as a fictional artifact in a video game. All of these instantiations of an invention contribute to its emergence. To assist student exploration and/or curation of collections, Fabric uses an open access content management system and presentation software, CollectiveAccess that draws on the Dublin Core™ Metadata standard to organize and standardize its fields of information. The researcher contribution interface facilitates adding invention artifacts and a public web interface provides means to identify, collect, archive, catalogue, revise and analyze the discourses (i.e., articles, images, audios, videos, other artifacts and events) surrounding emerging technologies. Whether examining, contributing, or curating Fabric collections, students become exposed to multiple stakeholders in the exploration of technology emergence through these different media genres.

(Duin et al., 2021, p. 96)

Beginning in 2019 and continuing to date, US and international scholar-instructors use Fabric as a learning database for development of digital literacy. Students engage with Fabric in one or more of these ways as detailed in Figure 6.1 below:

- Examine: Students explore the objects in a collection, examining their origin, feature, and potential uses in the society. Instructors may ask students to consider the rhetorical, social, or technical implications of these objects as part of the examination.
- Contribute: Students archive single objects using existing keywords and metadata on Fabric. Students learn to use media editing tools like image and video editors to create a thumbnail for the archived object.
- Curate: Students envision, create, and submit a new collection for possible publication at Fabric based on a thesis or unique point of view. Students identify and propose artifacts from within and outside of Fabric,

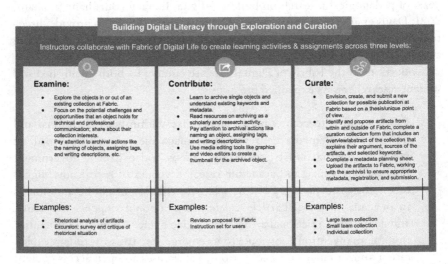

FIGURE 6.1 Building Digital Literacy (BDL) assignment types; from Davis et al. (2021). Used with co-author permission.

completing a curation collection form that includes an overview/abstract of the collection. To do this, they complete a Google (or Excel) sheet for metadata planning associated with each artifact, and they compose an overview of their collection.... Working with the archivist to ensure appropriate metadata, registration, and submission, they then upload their artifacts to Fabric through a customized interface. On the public interface, each item includes a thumbnail for the artifact housed at Fabric as well as the metadata.... Their metadata is then published as part of each artifact in their collection. (Duin et al., 2021, p. 97).

Specifically, scholar-instructors from organizations aligned with technical and professional communication (CPTSC, Council for Programs in Technical and Scientific Communication; ATTW, Association of Teachers of Technical Writing) have been invited to collaborate since 2019, meeting multiple times a term to explore Fabric, share assignment development, and articulate evolving definitions and work to cultivate student digital literacy. The title of the overall project is Building Digital Literacy (BDL), and its associated research cluster is officially part of the Digital Life Institute (https://www.digitallife.org/) housed at Ontario Tech University in Canada.

The following summaries of four publications illustrate collaborative work on assignment creation, deployment, and analysis of the ways that instructors and students make sense of their use of Fabric to build digital literacy. These studies represent a focus on sustainable pedagogical research as described by Lisa Melonçon et al. (2019) for the design of "thoughtful studies" and the need to perform multi-institutional research: "This research is difficult, complex, and cumbersome, but it is especially needed as multi-institutional research and field-wide research offer TPC the most potential for affecting day-to-day practices" (p. 22).

First, Davis et al. (2021) chronicle development and use of assignments implemented in multiple writing courses to "engage students with the multiple literacies needed to be successful critical thinkers in their personal and professional digital lives." Common across the BDL project, collaborators struggled to identify an agreed-upon meaning for digital literacy in writing classrooms, discussing how pedagogical scholarship voices a need to understand literacy as *literacies*, or as plural, multidimensional, or multilayered (Breuch, 2002; Cargile Cook, 2002; Selber, 2004; Hovde & Renguette, 2017). This pedagogical article provides direction for development of "examine" assignments in which students reflect critically on the rhetorical nature of digital artifacts/emerging technologies. In one assignment, students explore how curators construct arguments about a set of emerging technologies, reporting in memo format on the inclusion or exclusion of certain artifacts. Students can become more involved with Fabric as a digital archive through "contribute" assignments in which they locate new artifacts to add to a collection(s), name each artifact, assign tags, and write artifact descriptions. The most intense work comes with "curate" assignments in which students develop a new collection for possible publication on Fabric. The act of individually or collaboratively curating and archiving a collection involves student development of metadata and diving into new, unfamiliar technical tasks.

Kenyan Burnham and Jason Tham's (2021) publication, "Developing Digital Literacy through Multi-Institution Collaboration and Technology Partnership," went the next step through analysis of assignments, student responses, and instructor reflections. They investigated how TPC instructors "teach" digital literacy using the Fabric platform along with studying how meaningful the assignments were to students and instructors. Similar to Davis et al., they found that instructors collaboratively designed assignments that took advantage of the rhetorical nature of Fabric as a digital archive. Students in this study report overall positive experiences with the "examine," "contribute," and "curate" assignments, articulating their digital literacy development through sharing about the challenges encountered and how they overcome them. They wrote that:

> however challenging new technologies might be to understand and master, students emphasized social, rhetorical, and cultural challenges above technical challenges in relation to digital literacy learning, reminding TPC educators that the site of engagement for digital literacy development extends beyond the platform to the more complex paradigm that platforms like *Fabric* were situated within.

Furthermore, they once again emphasized that "we need to let students experiment with different technologies so they can cultivate new skills and literacies."

Also published in 2021, Jason Tham and colleagues share results from a grounded theory analysis to identify the ways students make sense of their use of Fabric and evolving understandings of digital literacy. Based on survey results, they found that students explicitly engaged prior knowledge (mental models) and metaphors in learning a new tool. This work continues to inform our developing framework for building digital literacy. Specifically, this study found that students employ tangible and intangible references like working a puzzle vs. formulating a story as a way of understanding new technology. Metaphors can be appropriate for sanctioning actions and helping students set goals as they orient themselves to new interfaces. Most of all, this study reinforces the notion that teaching digital literacy requires an approach that widens rather than narrows students' engagement with digital texts. Because digital literacy relies on multiple models and metaphors to connect the novel to the known, we should encourage students to make as many different connections as possible from their own experience to understand augmentation technologies. They conclude:

This study forwards the argument that understanding of digital literacy requires a broad approach that takes into account the many experiences learners bring to their use of digital tools. While we do not replace any existing frameworks or definitions for digital literacy—as they must consider the local contexts in which the frameworks or definitions are deployed—we learned from our study that student perceptions of their own digital literacy (and instructor's awareness of such) are informed by prior experiences by means of metaphors and mental models. These conceptualizations can shape how they learn with technology.

(Tham et al., 2021, pp. 14–15)

Throughout the BDL project and these instructional deployments, we continue to be struck by how building digital literacy is informed and influenced by prior experiences.

Last, the most recent study by Davis et al. (2022) examines the writing infrastructure of the Fabric platform to explore how, through these above assignments, students make visible the invisible layers of infrastructure. As a digital platform, Fabric's infrastructure archives, studies, and illuminates other infrastructures. It acts as infrastructure by providing a virtual platform for artifacts about technologies, promoting certain discourses or ways of seeing technologies through its collections, and providing opportunities to participate in building the platform, its collections, and its artifacts. In turn, the BDL research group incorporates Fabric into TPC classrooms by having students analyze, create, or contribute to collections. Specifically, Davis et al. used Rebecca Walton et al.'s (2019) 3Ps framework to illustrate how the above pedagogical approaches encourage students to use writing to interrogate digital infrastructure and the ways it is entangled with positionality, privilege, and power. This team analyzed assignment prompts, published collection descriptions created and contributed by students, and assembled collaborative autoethnographic reflections of this work. Through such curation work, students occupy a certain space of privilege and exert a level of power through rhetorical choices about how to represent technologies in collections. Analyses reveal how Fabric collections act as infrastructure by arranging specific narratives about technologies, narratives that may perpetuate reductive or harmful thinking. Fabric itself is an infrastructure that holds within it the power to contribute to social justice work within the TPC discipline, a means to encourage critical awareness of infrastructure. And again:

A common thread in many of the instructor reflections in our dataset was the notion that society is a multilayered, multidimensional network of actors and technologies with varying degrees of agency, and that interacting in such an infrastructure is messy, wavering, and requires complex digital literacies. As instructors try to help students develop digital literacy skills, it is clear that we constantly reflect on our processes and grapple with knowing that digital literacy is a privileged skill.

(Davis et al., 2022)

Use of Fabric across TPC and writing courses provides the opportunity to share agency with students, to "lay bare the rhetorical, agentive nature of infrastructure," to provide opportunity for greater understanding of metadata and how search algorithms shape engagement with information, knowledge making, and digital literacy. Ultimately, this analysis showed that use of Fabric in these classrooms contributed well beyond digital literacy:

> Through engagement with and metacognition about metadata, roles, and the ways that digital infrastructure is developed and influences our lives, we engaged students in thinking critically about infrastructure itself. We made visible the invisible functions of writing as creator, curator, and user of digital technologies. Further, we engaged students in thinking about social justice through consideration of the ways that infrastructure and positionality, privilege, and power are interconnected and mutually constitutive.
>
> *(Davis et al., 2022)*

These studies provide pedagogical direction for incorporating Fabric across the curriculum to expand knowledge of emerging technologies and build digital and AI literacy. As detailed in Chapter 4, Duri Long and Brian Magerko (2020) define AI literacy as "a set of competencies that enables individuals to critically evaluate AI technologies; communicate and collaborate effectively with AI; and use AI as a tool online, at home, and in the workplace" (p. 2). Table 6.1 is a modification of Table 4.1, again including Long and Magerko's four research questions and AI competencies. The last column includes ways to use the Fabric of Digital Life as a means to build understanding of augmentation technologies and AI literacy.

To date, Fabric includes a large number of innovative collections created by student teams. In addition to building digital and AI literacy, this work promotes student reconceptualization of augmentation technologies through a human-centric, ethical design framework as they acknowledge rich assemblages of emerging technologies. Organized by course, Table 6.2 includes course titles and associated collections developed by undergraduate and graduate students across TPC courses at multiple institutions. We encourage readers to access these collections (see the Fabric URLs in the third column); each Fabric collection page includes metadata specific to the artifacts' relation to the body as well as to what the artifacts augment (see discussion of metadata development in Chapter 2).

The focus throughout development of the above collections was primarily on building digital literacy; instructors did not necessarily discuss the concept

TABLE 6.1 Long and Magerko's (2020) research questions and AI competencies; Related use of fabric of digital life for developing augmentation technology and AI literacy

	AI competencies	Use of fabric
What is AI?	Recognizing and understanding AI; Interdisciplinarity	Focus on examining augmentation technologies at Fabric: Does it use AI? If so, how?
What can AI do?	Determining AI strengths and weaknesses; Imagined futures	Focus on examining augmentation technologies at Fabric that use AI: What strengths and weaknesses are present? How are futures being imagined?
How does AI work?	Understanding digital and data literacy; Critical interpretation	Focus on Fabric platforms; focus on identifying metadata: How is each technology explained and contextualized?
How do people perceive AI?	Documenting and influencing explainability	Focus on Fabric curation; identification of new artifacts, metadata literacy development and study: How do I determine metadata, and how does this work to build trustworthy AI?

of augmentation technologies. However, upon revisiting these collections, we see that each collection includes technologies that provide for cognitive, emotional, sensory, and/or physical enhancement. For example, note the focus on physical enhancement as part of the collection titled *Implanted and Embedded Medical Devices*, for which the student curators write this introduction:

> Our collection introduces the newest advances in implanted and embedded medical devices, spanning from the brain to the ankle. These devices are used to replace, repair, or enhance organ function and limbs. Some of the devices, such as the leadless pacemaker, have already been approved by regulatory agencies and are currently being used. Other devices are prototypes that are still being developed, such as the artificial kidney. Many of the devices are developed by large, multinational companies such as Medtronic, which impact health care across the globe.
>
> Some of these treatment options build upon old technologies, such as improving the pacemaker to remove the weakest elements (the leads) and improving battery life and monitoring capabilities. Some of these treatment options incorporate technologies such as augmented reality and 3D printing. Augmented reality was used to supplement vision in the bionic eye. 3D printing can be used to create personalized organs such as ankle bones that fit a particular person.

TABLE 6.2 Fabric collections developed by undergraduate and graduate students across TPC courses at multiple institutions

Course title	Collection title	Type of enhancement	Fabric URL
Undergraduate Technical and Professional Writing	Emerging Technologies for Technical Communication	Sensory & physical enhancement; Note discussion of machine learning and AI for cognitive enhancement	https:// fabricofdigitallife. com/Browse/ objects/facets/ collection:29 Tham (2019)
Undergraduate Technical and Professional Writing	Wearables and Carryables for Everyday Communication: Past, Present, Future	Sensory enhancement	https:// fabricofdigitallife. com/Browse/ objects/facets/ collection:35 Stambler (2019)
Upper-level undergraduate Technical Writing courses	Digital Translation Technologies for Healthcare Settings	Sensory enhancement	https:// fabricofdigitallife. com/Browse/ objects/facets/ collection:58 Clarke (2020)
Upper-level undergraduate Technical Writing courses	Emerging Audio-based Platforms and Participant Communities	Sensory enhancement	https:// fabricofdigitallife. com/Browse/ objects/facets/ collection:60 Gerace (2020)
Upper-level undergraduate Technical Writing courses	The Therapist in My Pocket: Mental Health Conversational Agents	Emotional enhancement	https:// fabricofdigitallife. com/Browse/ objects/facets/ collection:59 Bennett (2020)
Undergraduate Business and Professional writing	Emerging Technologies for Business Communication	Cognitive & sensory enhancement	https:// fabricofdigitallife. com/Browse/ objects/facets/ collection:42 Veeramoothoo (2019)
International Professional Communication (undergrad and grad)	What Language Sounds Like: Wearable Devices in Translation Communication	Sensory enhancement	https:// fabricofdigitallife. com/Browse/ objects/facets/ collection:31 Ryan et al. (2019)

Course title	Collection title	Type of enhancement	Fabric URL
International Professional Communication (undergrad and grad)	Cultural Reality—A VR Experience	Emotional & sensory enhancement	https://fabricofdigitallife.com/Browse/objects/facets/collection:32 Snyder et al. (2019)
International Professional Communication (undergrad and grad)	AR From Conception to Reality	Sensory enhancement	https://fabricofdigitallife.com/Browse/objects/facets/collection:33 Blissenbach et al. (2019)
International Professional Communication (undergrad and grad)	Implanted and Embedded Medical Devices	Sensory & physical enhancement	https://fabricofdigitallife.com/Browse/objects/facets/collection:34 Gruber et al. (2019)
International Professional Communication (undergrad and grad)	Digital Health Devices and Strategies	Sensory & physical enhancement	https://fabricofdigitallife.com/Browse/objects/facets/collection:51 Melby et al. (2020)
International Professional Communication (undergrad and grad)	Fostering a Culture of Transcreation for Improving Mistranslation and Miscommunication	Sensory & physical enhancement	https://fabricofdigitallife.com/Browse/objects/facets/collection:50 Burnes et al. (2020)
International Professional Communication (undergrad and grad)	Challenges with Improving Workplace Communication	Cognitive, sensory & emotional enhancement	https://fabricofdigitallife.com/Browse/objects/facets/collection:56 Anders et al. (2020)
International Professional Communication (undergrad and grad)	Japanese Technologies	Cognitive, physical & sensory enhancement	https://fabricofdigitallife.com/Browse/objects/facets/collection:52 Campbell et al. (2020)

(*Continued*)

Course title	Collection title	Type of enhancement	Fabric URL
International Professional Communication (undergrad and grad)	Using Technology to Navigate Foreign Lands and Cultures	Sensory enhancement	https:// fabricofdigitallife. com/Browse/ objects/facets/ collection:54 Sparrow et al. (2020)
Business and Professional Communication (grad)	Replacing Human Decision Making Through Automated Tracking Systems (ATS) in Hiring Processes	Cognitive enhancement	https:// fabricofdigitallife. com/Browse/ objects/facets/ collection:62 Hocutt et al. (2020)
Writing with Digital Technologies (grad)	Non-Traditional Prosthetics	Physical and cognitive enhancement	https:// fabricofdigitallife. com/Browse/ objects/facets/ collection:37 Molohon et al. (2019)
Writing with Digital Technologies (grad)	Biotechnology and Human Health: Harvesting the Technology of Plants and Microbes to Augment the Human Body	Sensory & physical enhancement	https:// fabricofdigitallife. com/Browse/ objects/facets/ collection:39 De La Victoria et al. (2019)
Writing with Digital Technologies (grad)	Surveillance, Sousveillance, and Security Technologies: A Variety of Wearable Computing Devices	Physical, sensory & cognitive enhancement	https:// fabricofdigitallife. com/Browse/ objects/facets/ collection:38 Banks & Burnes (2019)
Writing with Digital Technologies (grad)	VR as a Sales Tactic	Physical & sensory enhancement	https:// fabricofdigitallife. com/Browse/ objects/facets/ collection:40 Campbell et al. (2019)

The themes of "innovation" and "hope" surround all of these different medical technologies. These medical devices can offer hope for patients living with conditions that have not responded well to lifestyle adjustments and medication. For some, these devices have led to a significant increase in their quality of life. Announcements of new devices tend to have a strong focus on their positive potential. However, there is always a trial-and-error process with new technologies even after they are approved. We have also included a database of international medical devices where users can search for recalls or adverse effects of a specific device.

(Gruber et al., 2019)

Another example, in this case for sensory enhancement, is the collection titled *What Language Sounds Like: Wearable Devices in Translation Communication*, in which student curators ask, "How do these technologies affect the other nuances like traditions and cultural behaviors that to date have been such a vital part of intercultural communications? Is raw translation enough? How important are the other cultural dimensions?" (Ryan et al., 2019).

This hands–on experience of examining and/or curating emerging technologies via use of the Fabric of Digital Life positions students as researchers who share their "productive ambiguity" when completing these assignments. As Davis et al. (2021) write:

As we reflected on our assignment deployments and student reflections, it was clear that difficulties and outcomes were closely intertwined— what we identified as challenges also spurred successes in some cases. Our reflective discussions overwhelmingly pointed to a *productive ambiguity* that students felt in working with these assignments. We found that our students' struggles to navigate the critical layers of digital literacy simultaneously resulted in some successes—the 'productive' element of our 'productive ambiguity.'

In short, the Fabric of Digital Life archive provides an innovative opportunity for instructors to design and integrate literacy development and increased understanding of augmentation technologies and AI that caters to student professional futures and workforce preparedness. While these examples represent work in individual courses, we turn next to the need to position expanded development of augmentation and AI literacies across TPC curricula.

Positioning Development of Literacies across TPC Curricula

Amid increased emergence of new technical communication programs, in 2005, Sandi Harner and Anne Rich worked to determine the most commonly required technical communication courses across the then 80 bachelor's degree programs in the US to articulate core skills and the identity of this field. The top five required or elective courses across these programs included technical communication, technical editing, advanced technical communication, visual communication, and production tools. In some cases, the programs included production courses oriented toward online and user documentation and information design. Envisioning use of Fabric, each of these courses could well include instructional modules of the type described in the previous section.

As a means to update Harner and Rich's work, in 2013 Lisa Melonçon and Sally Henschel located 185 undergraduate TPC programs, restricting their curricular analysis to the 65 programs that offered majors. They identified the following core courses: technical communication, technical writing, senior portfolio, capstone, editing, internships, introductory courses, web production, document design, and genre courses. Comparing the two studies, Melonçon and Henschel describe curricular change from 2003 to 2011 as "a period of tremendous technological and communication change" (p. 54). Again, instructors could easily embed use of Fabric in each of these courses to build digital and AI literacy.

Since 2013, there has been increased focus on learning outcomes by accrediting agencies, institutions, and programs. In 2021, Geoffrey Clegg, Jessica Lauer, Johanna Phelps, and Lisa Melonçon coded and analyzed data from 376 programmatic student learning outcomes (PSLOs) from 47 TPC undergraduate programs. As they state, PSLOs "focus on practical and conceptual skills and indicate what students are expected to learn through curricular and co-curricular activities associated with programs" (p. 19). Their analysis determined four outcomes coded as *Writing, Rhetoric, Technology,* and *Design* as being consistent across at least half of the institutions (p. 24). For purposes of this chapter, we share their findings related to *Technology* "as a code to capture the field's programmatic emphasis on technology... view[ed] as both a tool and a cultural force," with example outcomes as follows:

- Demonstrate a critical perspective of technology, its uses, users, and contexts
- Students will use a variety of communication tools
- Students will have the ability to use, analyze, and learn communication technologies (p. 25)

Quoting Brumberger's (2003) earlier argument, they underscore the need for both understanding of tool and its cultural force:

> The field of technical communication is in many ways inscribed by technology. As a result, technical communication programs not only must provide students with a foundation in the theory and practice of

the field, but also must give students some level of proficiency in the technology tools they will need to put that knowledge into service in the workplace.

(p. 64)

Clegg and colleagues further emphasize that this dual emphasis:

is vital in preparing students to not only use the technology but to be able to make thoughtful, critical decisions about it. Without the dual emphasis—use and critique—TPC programs leave students vulnerable to either not being prepared to enter the workforce or not being prepared to make ethical choices in the workplace through critique and analysis.

(p. 26)

Also aligning with chapters here is their strong recommendation for programs to include *Ethics, Research, Collaboration,* and *Professionalization* as additional outcomes, and again, Fabric represents an innovative site for work on each of these outcomes as a means toward student work preparedness.

In their study about preparing students for roles as leaders and innovators amid such change, Carlos Evia and Rebekka Andersen (2018) interviewed academic and industry colleagues, and by so doing, determined key skills for "this new level of practice and innovation" that includes systems thinking, computational thinking, business analysis, and technical aptitude. They write that "to lead and innovate in technical communication, professionals now need these skills in addition to the more traditional skills in rhetorical analysis, writing, and design expected of technical communicators." A search for "systems thinking" alone brings up 650 artifacts at the Fabric of Digital Life, including the collection on *Surveillance, Sousveillance, and Security Technologies.* Again, we encourage use of Fabric for theoretical and pragmatic model development for instructor-scholars as a method to design and integrate literacy development that caters to student learning needs as well as their professional futures and workforce preparedness. For reflections by instructors who have done so, please see the recordings and documents as part of the scholar-instructor *Collaborative Autoethnography for Building Digital Literacy* collection at Fabric (Duin et al., 2020).

We also emphasize the need to position expanded development of augmentation and AI literacies across TPC curricula. As an example, we share the current curricula for the graduate certificate in technical communication and MS programs in Scientific and Technical Communication at the University of Minnesota, both housed in the Department of Writing Studies (WRIT). The core coursework for these programs is similar for both programs (15 cr); the MS degree includes an elective course (3 cr), two outside area courses (6 cr), and a capstone course (3 cr):

Core Coursework (15 credits)

WRIT 5001—Introduction to Graduate Studies in Scientific and Technical Communication (3.0 cr)

WRIT 5112—Information Design: Theory and Practice (3.0 cr)

WRIT 5501—Usability and Human Factors in Technical Communication (3.0 cr)

WRIT 5561—Editing and Style for Technical Communicators (3.0 cr)

WRIT 5662—Writing With Digital Technologies (3.0 cr)

Capstone Course (3 credits)

WRIT 8505—Professional Practice (3.0 cr)

Electives (3 credits)

Students take at least 3 credits of electives, or in some cases, 3 credits of independent study (WRIT 5291) or internship (WRIT 5196).

WRIT 4562—International Professional Communication (3.0 cr)

WRIT 4573W—Writing Proposals and Grant Management (3.0 cr)

WRIT 5664—Science, Medical, and Health Writing (3.0 cr)

Development of augmentation and AI literacies, while certainly a key component of the Writing with Digital Technologies course, might best be integrated across this curriculum. For example, students can be introduced to the four augmentation technology dimensions as part of the introduction to this field; they can test out AI-writing tools and techniques as part of both the Information Design and Editing and Style courses; they can work with explainable AI in Usability and Human Factors; and all of the above courses offer the opportunity to use Fabric of Digital Life as a critical resource throughout their academic and professional careers. Moreover, students might even create a digital student/ employee to be paired with throughout this curriculum (see Chapter 7).

As an undergraduate curriculum example, the following are the courses required for the Technical Writing and Communication BS program at the University of Minnesota:

Foundation Course

WRIT 1001—The Art of Explaining Things: Introduction to Technical Writing and Communication (3.0 cr)

Core Courses (7 required courses, 22 cr, from the following list)

WRIT 3221—Communication Modes and Methods (3.0 cr)

WRIT 3441—Editing, Critique, and Style (3.0 cr)

WRIT 3671—Visual Rhetoric and Document Design (3.0 cr)
WRIT 3701—Rhetorical Theory for Writing Studies (3.0 cr)
WRIT 4501—Usability and Human Factors in Technical Communication (3.0 cr)
WRIT 4662—Writing With Digital Technologies (3.0 cr)
WRIT 3562—Technical and Professional Writing (4.0 cr)
or WRIT 3562—Honors: Technical and Professional Writing (4.0 cr)

For this BS program, students take four additional elective courses from Writing Studies (12 cr), a capstone course (3 cr), and 15 credits from outside departments whose courses assist students in completing one of four sub-plans: information technology and design, biological and health sciences, legal discourse and public policy, or environmental science. Similar to the professional graduate programs in technical communication, development of augmentation technology and AI literacies, while certainly a key component of the Writing with Digital Technologies course, again might best be integrated across this curriculum. Students can be introduced to the four augmentation technology dimensions as part of the foundation course for this field; they can test out AI-writing tools and techniques as part of the editing and visual rhetoric/document design courses; they can work with explainable AI in Usability and Human Factors; and all of the above courses offer the opportunity to use Fabric of Digital Life as a critical resource throughout their academic and professional careers.

Augmentation Technology Literacies

This book challenges scholars, instructors, and practitioners to reframe practice to promote literacies surrounding the ethical design, adoption, and adaptation of augmentation technologies. Here we revisit the four augmentation technology dimensions—cognitive, sensory, emotional, and physical enhancements—with attention to pedagogical direction for building multiple literacies.

When you encounter an augmentation technology or begin teaching this concept, we encourage you to begin by asking, "How does the proof-of-concept or product promise claim to augment a human?"

- Cognitive enhancement:
  - Does it help someone to be smarter, more knowledgeable, think faster, remember more, know more, learn faster, learn more efficiently, edit dysfunctional memory, or be more reasonable?
  - Does it enhance human intelligence, or is the device being promoted to replace it?
- Sensory enhancement:

- Does it help someone to experience more through augmentation of the senses, to augment or reconstruct reality through digital representations, to escape one's sensory reality, or to focus better?
- What promises are being made to advance human abilities significantly for increased efficiency and automation?
- Emotional enhancement:
  - What ambitious goals are being set for use of emotional enhancement as a means to address grand challenges (e.g., care systems for aging populations)?
  - Does it help someone to understand one's emotions, to control one's emotion, or even be happier or more fulfilled?
- Physical enhancement:
  - Does it help someone to be stronger, faster, live longer, live healthier, work longer, physically work more efficiently, or play sports better?
  - Is the device based on legitimate science? Does it celebrate superheroism? Does it react to a shared sense of vulnerability or fear of the future?

As an example, consider Amelia, a conversational AI framed as a "digital employee," a term increasingly used in industry (see Chapter 7). Amelia uses natural language processing, speech recognition and speech generation, and other AI technologies. The videos at https://amelia.ai promise that Amelia *can work autonomously for you*, making her appear to be a significant upgrade to a Siri or Google Assistant. The marketing copy consistently promotes autonomous work, rather than human supervision through the Amelia Integrated Platform (Figure 6.2).

- How does Amelia enhance human intelligence, or is the device being promoted to replace it?
- How might Amelia help someone to be smarter, more knowledgeable, think faster, remember more, know more, learn faster, or learn more efficiently?
- Might Amelia help someone to understand or control emotion?

At this juncture, instructors, practitioners, and students should become familiar with the broad range of autonomous tools currently available. In *Writing Futures* and later as part of conference presentations (Duin & Pedersen, 2021, 2022), we encouraged readers to explore these examples:

- The Google Deep Dream project reignites public curiosity about AI-generated art and its psychedelic reproductions of patterns within patterns.
- Microsoft's Xiaolce chatbot, a Chinese language conversational AI, converts images into Chinese poetry.

FIGURE 6.2 Amelia, an example of conversational AI. Used with permission of Amelia US LLC. https://amelia.ai/conversational-ai/.

- Manuscript Writer by SciNote (2022) empowers scientists by pulling information from data in SciNote, keywords, and DOI numbers of open access references. It then presents all of this as a complete manuscript.
- Novelist Sigal Samuel describes how her use of GPT-2 (Generative Pretrained Transformer, a neural network machine learning model) helps her to see things anew, and Vincent discusses how the system was trained on 8 million text documents scraped from the web. Also using GPT-2, EssaySoft AI generates academic essays.
- Its next iteration, GPT-3 generates thousands of articles and books in one day by:

> learning how to write by reading millions of articles and books from Google Books and other websites. It can analyze any piece of text along with an existing article or book in seconds. It then captures the style of writing for all new articles that users want to write.
>
> *(Ravichandran, 2022)*

A video by AI Content Dojo examines three methods—blank canvas, forms-based, and mixed—for use of GPT-3 AI-writing applications.

- Techradar.com writers Anna Sevilla and Mike Jennings (2022) highlight the "best AI writer" tools each year. These include Writesonic (using GPT-3) for creation of ads, blogs, landing pages, and product descriptions; Article Forge, also for website content development, including text in seven languages for multiple markets; WordAI for revisions of original articles for greater readability, quality, and clarity; and AI Writer for generating articles from scratch or revisions.
- Kate Knibbs (2022) writes about how a novelist and an AI co-wrote the new novel, *Amor Cringe*, billing itself as "deepfake autofiction."
- Jasper, the Future of Writing at https://www.jasper.ai, is AI for creating blog, social media, and website content. Alan Knowles (2022) examines the impact of AI (Jasper) on the practice of writing digital marketing copy, finding that Jasper functions "more like a professional writing collaborator than… a mere writing tool" (p. 259).

OpenAI's ChatGPT was released in late 2022, with thousands of articles now written about its use across education and professions. Millions are using it to write, create, and edit numerous forms of documents. Microsoft is poised to spend $10B to acquire OpenAI's ChatGPT technology to power its search engine Bing and be incorporated into its suite of tools. Other creative AI tools stemming from GPT-3 will continue to be released, significantly transforming communicative actions and digital behaviors, with GPT-4 on the horizon. Google also shared plans for release of its conversational AI chatbot, Bard, powered by its Language Model for Dialogue Applications (LaMDA). These examples demonstrate the power of large language models and generative AI, with broad implications for writing and search tools.

Scholars and practitioners are designing and studying the pedagogy surrounding the use of AI writers in technical communication (Anson & Straume, 2022), discussing new modes of learning enabled by AI (Mollick & Mollick, 2022), dissecting angst regarding academic integrity (Marche, 2022), and providing direction for ChatGPT use throughout the user testing process (e.g., see Pybus, 2023; Tisza, 2022).

As part of one's exploration and experimentation with these AI augmentation technologies, recall the ethical themes, principles, and consequences discussed in Chapter 5. Moreover, all of the above links position instructors and students to use AI to learn independently of program curricula and to continue as lifelong learners amid a world of AI augmentation technologies. As Lina Markauskaite and her 10 colleagues (2022) wrote upon engaging in "a semi-independent and semi-joint written polylogue" on rethinking the entwinement of AI and human learning:

> There is little doubt that AI reconfigures the distribution of intelligence, labour and power between humans and machines, and thus new kinds of capabilities are needed (Luengo-Oroz et al., 2020). However, these capabilities are as yet poorly identified and understood.
>
> *(p. 2)*

They ask:

> What kind of capabilities do people need for successful cooperation, functioning, and wellbeing in an interconnected and fast-changing world permeated with AI? How can we conceptualise these capabilities? How can we help learners develop them? How can we empirically study and assess their development?.
>
> *(p. 2)*

They adopt the term "capacity" and "capabilities" to shift focus from demonstrated behaviors to "the potential, dispositions and opportunities within one's reach to pursue specific values and outcomes" (p. 2). The return to use of "capabilities" aligns with the UK's Joint Information Systems Committee (Jisc) Digital Capability Framework (Jisc, 2022) discussed earlier. However, Markauskaite et al. emphasize that "it is essential to consider new possibilities and barriers one might encounter in enacting and enhancing those capabilities that now become distributed between humans and intelligent machines when humans and machines perform in cooperation" (p. 2).

In Chapter 5 we defined and discussed international AI socio-ethical principles and envisioned them for augmentation technologies. Here we present the findings of the Markauskaite et al. stakeholder group as they strongly parallel Chapter 5's guidance. Their collective results provide critical pedagogical direction (note each co-author's specific contribution):

1   Become an "agentic learner"
    To maintain job relevance and support future career transitions in a world with AI, individuals will require highly developed self-regulated learning (SRL) skills (Winne et al., 2017).... The need for SRL skills is even more acutely emphasized in the age of AI due to two prominent reasons: (i) the need to adapt (re- or up-skill) frequently due to speed of job and life changes; and (ii) the need to maintain agency in decision making while working AI systems (co-author Dragan Gašević, p. 3).

2   Develop "broad intelligence"
    To interpret and understand AI system outputs, [individuals should] integrate these outputs into human knowledge systems, assess and evaluate ethical implications of AI outputs, and elevate human cognitive work to creativity and meaning/sense-making domains (co-author George Siemens, p. 4).

3   Foster human creativity
    AI can support creativity, particularly Pro C and potentially Big C, as it can extend an expert's knowledge. Yet, it does not replace mini c or little c creativity. At the mini c and little c levels, the creative output is not as important as the self-discovery that occurs through the creative process. It is therefore important to develop both an appreciation and understanding of when and where AI is most useful (co-author Rebecca Marrone, p. 5).

4   Empower people to make free choices about AI; cultivate "deliberate engagement" with AI
    To engage mindfully demands that individuals understand how AI technology operates. Notably, this technical knowledge is inseparable from the availability of systemic freedom: people need to have the choice between 'effects of' or 'effects with' to achieve what they value at a given moment (co-author Oleksandra Poquet, p. 6).

5   Create AI for human values
    Those capabilities related to design, ethics and philosophy will be critical for creating AI for human values. Design thinking and design skills will enable humans to decide the degree of human control and computer automation for solving key human challenges. Helping people to develop ethical and philosophical thinking from early stages of education, will enable the future designers and developers of AI to create innovations that keep the human values at the center (co-author Roberto Martinez-Maldonado, p. 7).

6   See self and AI in a larger system
    What are the competencies teachers need to function in an AI world and to
    prepare their students for an AI world?... A key capability of teachers will
    be to critically engage with new technologies and consider them in rela-
    tion to learning experiences and their own teacher work (cf. JISC, 2019).
    This requires teachers to have some level of data literacy, specifically how
    to make decisions about which tools to use and for what purposes. [And
    institutions will need to provide] the necessary support to engage with
    AI-enabled digital technologies and data (co-authors Sarah Howard and
    Jo Tondeur, p. 7).
7   Navigate AI-mediated world views
    We should focus on capabilities (and learning) to allow learners to un-
    derstand the world around them—including the role of AI in it—but also
    to understand their role in reshaping that world; learners, technologies,
    and society are intertwined or mutually constitutive in this regard. The
    skills and knowledge developed around coding, data literacy, understand-
    ing how algorithms and other technologies work, are all crucial not just
    for workforce participation, but for civic participation (co-author Simon
    Knight, p. 8).
8   Foster representational literacy for collective sense-making
    The proposition then, is that the ability to use interactive visualizations in
    ways that invite the construction of a shared narrative with stakeholders
    in the room (physical or online) can be learnt, practiced and improved....
    First, we must become more fluent in judging if/when to bring AI agents
    into the conversation to help move deliberations forward. Second, the vi-
    sual representations we use to express ideas are increasingly interpretable
    computationally. Intelligent agents can now become part of the conversa-
    tion, even contributing their own issues, ideas and arguments (e.g., Buck-
    ingham Shum et al., 2010) (co-author Simon Buckingham Shum, p. 9).
9   Learn and create value in networks of humans and non-humans
    Professionals need capabilities to learn in the networks of humans and
    non-human intelligent systems. Increased digitalization, combined with
    advancement in learning analytics and AI opens the possibility for design-
    ing automated real-time feedback systems capable of just-in-time, just-in-
    place support for learning, complex problem solving and decision making
    (De Laat et al., 2020). These systems can augment learning and professional
    development in situ and can impact human and machine collaboration
    during teamwork (Seeber et al., 2020) (co-author Maarten De Laat, p. 10).

From their synthesis of these combined directions, Markauskaite et al. provide
AI competencies, which, in alignment with this book, focus on humanistic,
cognitive, and social perspectives.

AI indeed reconfigures the distribution of intelligence, labor, and power between humans and machines, and increased capabilities (augmentation technology literacies) are needed that center on humanistic, cognitive, and social perspectives. The development of these capabilities (augmentation technology literacies) is both individual and collaborative. We concur with Markauskaite et al. that "to develop these capabilities in students, teachers must also possess those capabilities.... Therefore, the question is how AI changes the world around us and how we choose to engage with AI as part of that world" (p. 13).

Again, the purpose of this book is to understand how augmentation technologies and AI change the world—and technical and professional communication—around us and how we choose to engage with augmentation technologies and AI as part of that world. Our goal is to reframe professional practice and pedagogy to promote digital and AI literacy surrounding the ethical design, adoption, and adaptation of augmentation and AI technologies. In Chapter 4, we discussed core competencies and design considerations for building digital and AI literacies; this included focus on explainable AI and trustworthy AI. Chapter 5 explored relevant consequences for augmentation technologies and pointed to future designs and practices that may mitigate potential harms. Here in Chapter 6 we have shared pedagogical direction for cultivating digital literacy and AI literacy skills needed to assess integration and participation with augmentation technologies. The benefits and risks, consequences and impact for learners are different with augmentation technologies because the technologies are evolving so rapidly.

Contributing as part of Markauskaite et al. (2022), Maarten De Laat poses the key challenge as becoming conscious about "how humans engage in hybrid relationships with AI," "how technologies shape and are shaped by activity," and "how collaborative inquiry and joint action are achieved in the face of shared challenges" (p. 10). We agree. Understanding these elements and their part in shaping human-machine practices is paramount for the development and design of policies, structures and architectures to facilitate learning. Chapter 7 focuses on these elements, chronicling the development of digital tools as autonomous agents or digital employees.

To conclude, Rand Waltzman, adjunct senior information scientist at the Rand Corporation, shares a plausible scenario in the metaverse now under rapid development:

> A political candidate is giving a speech to millions of people. While each viewer thinks they are seeing the same version of the candidate, in virtual reality they are actually each seeing a slightly different version. For each and every viewer, the candidate's face has been subtly modified to resemble the viewer.

This is done by blending features of each viewer's face into the candidate's face. The viewers are unaware of any manipulation of the image. Yet they are strongly influenced by it: Each member of the audience is more favorably disposed to the candidate than they would have been without any digital manipulation…

At the heart of all deception is emotional manipulation. Virtual reality environments, such as Facebook's (now Meta's) metaverse, will enable psychological and emotional manipulation of its users at a level unimaginable in today's media.

(8 August 2022)

Amid this age of augmentation technologies and artificial intelligence, the interfaces TPCs design, the content displayed, and the navigation paths presented will be subtly and not so subtly modified to direct and assist the user. Digital employees (see Chapter 7) will assist across the design, development, adoption, and adaptation of technologies. While Waltzman writes that "we are not even close to being able to defend users against the threats posed by this coming new medium," we know that TPCs have always been and always will be positioned to defend, support, caution, and direct users. TPCs can be positioned to adroitly envision, develop, test, and deploy the "guardrails" that are now imperative for designing ethical futures. TPCs must pay serious attention to audience addressed, invoked, involved, immersed, augmented, and now ever more influenced.

References

ALA (2022). *Digital literacy*. American Library Association. https://literacy.ala.org/digital-literacy/

Anders, R., Smith, J., & Smrekar, K. (2020, June). *Challenges with improving workplace communication* [Multimedia collection]. Fabric of Digital Life. https://fabricofdigitallife.com/Browse/objects/facets/collection:56

Anson, C. M., & Staume, I. S. (2022). Amazement and trepidation: Implications of AI-based Natural Language Production for the teaching of writing. *Journal of Academic Writing, 12*(1), 1–9. http://dx.org/10.18552/joaw.v12i1.820

Banks, P., & Burnes, L. (2019, December). *Surveillance, sousveillance, and security technologies: A variety of wearable computing devices* [Multimedia collection]. Fabric of Digital Life. https://fabricofdigitallife.com/Browse/objects/facets/collection:38

Bennett, H. (2020, September). *The therapist in my pocket: Mental health conversational agents* [Multimedia collection]. Fabric of Digital Life. https://fabricofdigitallife.com/Browse/objects/facets/collection:59

Blissenbach, R., Brummer, N., Krstic, D., & Yoong, C. (2019, April). *AR from conception to reality* [Multimedia collection]. Fabric of Digital Life. https://fabricofdigitallife.com/Browse/objects/facets/collection:33

Breuch, L. K. (2002). Thinking critically about technological literacy: Developing a framework to guide computer pedagogy in technical communication. *Technical Communication Quarterly, 11*(3), 267–288.

Brumberger, E. (2003). *Beyond the borders of "English": Teaching technology tools in the undergraduate technical communication curriculum.* Paper presented at the council of programs in technical and scientific communication, Potsdam, NY.

Buckingham Shum, S., Sierhuis, M., Park, J., & Brown, M. (2010). Software agents in support of human argument mapping. In *Proceedings of third international conference on computational models of argument.* IOS Press.

Burnes, L., Fundingsland, S., Kosmalski, T., & Stromback, C. (2020, May). *Fostering a culture of transcreation for improving mistranslation and miscommunication* [Multimedia collection]. Fabric of Digital Life. https://fabricofdigitallife.com/Browse/objects/facets/collection:50

Burnham, K., & Tham, J. (2021). Developing digital literacy through multi-institution collaboration and technology partnership: An analysis of assignments, student responses, and instructor reflections. *Programmatic Perspectives, 12*(2). https://cptsc.org/wp-content/uploads/2021/12/Developing-Digital-Literacy.pdf

Byrd, A. (2020). 'Like coming home': African Americans tinkering and playing toward a computer code bootcamp. *College Composition and Communication, 71*(3), 426–452.

Campbell, A., Peterson, J., & Schaust, S. (2019, December). *VR as a sales tactic* [Multimedia collection]. Fabric of Digital Life. https://fabricofdigitallife.com/Browse/objects/facets/collection:40

Campbell, A., Pitsch, C., & Reinsel, D. (2020, June). *Japanese technologies* [Multimedia collection]. Fabric of Digital Life. https://fabricofdigitallife.com/Browse/objects/facets/collection:52

Cargile Cook, K. (2002). Layered literacies: A theoretical frame for technical communication pedagogy. *Technical Communication Quarterly, 11*(1), 5–29.

Clark, I. (1995). Information literacy and the writing center. *Computers and Composition, 12*(2), 203–209.

Clarke, M. (2020, June). *Digital translation technologies for healthcare settings* [Multimedia collection]. Fabric of Digital Life. https://fabricofdigitallife.com/Browse/objects/facets/collection:58

Clegg, G., Lauer, J. Phelps, J., & Melonçon, L. (2021). Programmatic outcomes in undergraduate technical and professional communication programs. *Technical Communication Quarterly, 30*(1), 19–33.

Dainton, B., Slocombe, W., & Tanyi, A. (Eds.). (2021). *Minding the future: Artificial intelligence, philosophical visions and science fiction.* Springer Publishing.

Davis, K., Stambler, D. M., Campbell, J. L., Hocutt, D. L., Duin, A. H., & Pedersen, I. (2022). Writing infrastructure with the Fabric of Digital Life platform. *Communication and Design Quarterly, 10*(2), 44–56.

Davis, K., Stambler, D., Veeramoothoo, C., Ranade, N., Hocutt, D., Tham, J., Misak, J., Duin, A.H., & Pedersen, I. (2021). Fostering student digital literacy through the Fabric of Digital Life. *Journal of Interactive Technology and Pedagogy.* Retrieved from https://jitp.commons.gc.cuny.edu/fostering-student-digital-literacy-through-the-fabric-of-digital-life/

De La Victoria, M., Finnoff, L., Fundingsland, S., & Kosmalski, T. (2019, December). *Biotechnology and human health: Harvesting the technology of plants and microbes to augment the human body* [Multimedia collection]. Fabric of Digital Life. https://fabricofdigitallife.com/Browse/objects/facets/collection:37

De Laat, M., Joksimovic, S., & Ifenthaler, D. (2020). Artificial intelligence, real-time feedback and workplace learning analytics to support in situ complex problem-solving: A commentary. *The International Journal of Information and Learning Technology, 37*(5), 267–277.

Duin, A. H., Caldwell, S., Campbell, J., Fonash, S., Hocutt, D., Ranade, N., Stambler, D., & Tham, J. (2020). *Collaborative autoethnography for building digital literacy* [Multimedia collection]. Fabric of Digital Life. https://fabricofdigitallife.com/Detail/collections/57

Duin, A. H., & Hansen, C. J. (Eds.). (1996). *Nonacademic writing: Social theory and technology.* Lawrence Erlbaum Associates.

Duin, A. H., & Pedersen, I. (2021). *Working alongside non-human agents.* IEEE International Professional Communication Conference (ProComm), October 2021.

Duin, A. H., & Pedersen, I. (2023). *Pedagogical direction for cultivating augmentation* Retrieved from https://fabricofdigitallife.com/Browse/objects/facets/collection:74

Duin, A. H., & Pedersen, I. (2022). *Tracing the turn to artificial human and human teaming.* IEEE International Professional Communication Conference (ProComm), July 2022, Limerick, Ireland.

Duin, A. H., Pedersen, I., & Tham, J. (2021). Building digital literacy through exploration and curation of emerging technologies: A networked learning collaborative. In N. B. Dohn, S. B. Hansen, J. J. Hansen, M. deLaat, & T. Ryberg (Eds.), *Conceptualizing and innovating education and work with networked learning* (pp. 93–114). Springer.

Duin, A. H., & Tham, J. (2018). Cultivating code literacy: A case study of course redesign through advisory board engagement. ACM SIGDOC, *Communication Design Quarterly, 6*(3), 44–58.

Evia, C., & Andersen, R. (2018). Preparing the next generation of leaders and innovators in technical communication. *Intercom,* July-August, 23–24.

Gerace, E. (2020, September). *Emerging audio-based platforms and participant communities* [Multimedia collection]. Fabric of Digital Life. https://fabricofdigitallife.com/Browse/objects/facets/collection:59

Gruber, K., Gawtry, S., & Peterson, J. (2019, April). *Implanted and embedded medical devices* [Multimedia collection]. Fabric of Digital Life. https://fabricofdigitallife.com/Browse/objects/facets/collection:34

Gurak, L. J. (2001). *Cyberliteracy: Navigating the internet with awareness.* Yale University Press.

Haas, C. (1996). *Writing technology: Studies on the materiality of literacy.* Taylor & Francis.

Harner, S., & Rich, A. (2005). Trends in undergraduate curriculum in scientific and technical communication programs. *Technical Communication, 52*(2), 209–220.

Hocutt, D., Baugham, H., Beck, S., Ciufo, R., Duncan, E., Gaitley, J., Gormus, S., Khoury, M., Lee, D., McClain, T., Ostrick, M., Plisko, M., Polivka, T., Rodriguez, G., Rucci, J., Rusbuldt, J., Sanderson, L., Tennyson, D., Vitkus, H., & Zlateva, D. (2021, January). *Replacing human decision-making through automated tracking systems (ATS) in hiring processes* [Multimedia collection]. Fabric of Digital Life. https://fabricofdigitallife.com/Browse/objects/facets/collection:62

Hovde, M. R., & Renguette, C. C. (2017). Technological literacy: A framework for teaching technical communication software tools. *Technical Communication Quarterly, 26*(4), 395–411.

Iliadis, A., Liao, T., Pedersen, I., & Han, J. (2021). Learning about metadata and machines: Teaching students using a novel structured database activity. *Journal of Communication Pedagogy, 4,* 152–165.

Jasper (2022). *Meet Jasper: The future of writing.* https://www.jasper.ai

JISC (2019). Teacher profile (higher education): Six elements of digital capabilities. https://digitalcapability.jisc.ac.uk/what-is-digital-capability/individual-digital-capabilities/

JISC (2022). *Building digital capability.* https://www.jisc.ac.uk/building-digital-capability

Knibbs, K. (2022, May 24). A novelist and an AI cowrote your next cringe-read. *Wired.* https://www.wired.com/story/k-allado-mcdowell-gpt-3-amor-cringe/

Knowles, A. (2022). Human-machine collaborative writing: Reduction of the human cognitive task load. In *Proceedings of the 2022 IEEE International Professional Communication Conference (ProComm),* July 17–20, 2022, Limerick, Ireland, pp. 257–261.

Long, D., & Magerko, B. (2020). What is AI literacy? Competencies and design considerations. In *CHI '20: Proceedings of the 2020 CHI Conference on Human Factors in Computing Systems,* April 25–30, 2020, Honolulu, HI, USA. https://doi.org/10.1145/3313831.3376727

Luengo-Oroz, M., Hoffmann Pham, K., Bullock, J., Kirkpatrick, R., Luccioni, A., & Rubel, S., et al. (2020). Artificial intelligence cooperation to support the global response to COVID-19. *Nature Machine Intelligence, 2,* 295–297. https://doi.org/10.1038/s42256-020-0184-3

Marche, S. (2022, December 6). The college essay is dead. *The Atlantic.* https://www.theatlantic.com/technology/archive/2022/12/chatgpt-ai-writing-college-student-essays/672371/

Markauskaite, L., Marrone, R., Poquet, O., Knight, S., Martinez-Maldonado, R., Howard, S., Tondeur, J., De Laat, M., Shum, S.B., Gašević, D., & Siemens, G. (2022). Rethinking the entwinement between artificial intelligence and human learning: What capabilities do learners need for a world with AI? *Computers and Education: Artificial Intelligence, 3,* 1–16. https://doi.org/10.1016/j.caeai.2022.100056

Martin, A., & Grudziecki, J. (2006). DigEuLit: Concepts and tools for digital literacy development. *Innovations in Teaching and Learning in Information and Computer Sciences, 5*(4), 249–267.

Melby, M., Khan, F., & Banks, P. (2020, May). *Digital health devices and strategies* [Multimedia collection]. Fabric of Digital Life. https://fabricofdigitallife.com/Browse/objects/facets/collection:51

Melonçon, L., & Henschel, S. (2013). Current state of U.S. undergraduate degree programs in technical and professional communication. *Technical Communication, 60*(1), 45–63.

Melonçon, L., Rosselot-Merritt, J., & St.Amant, K. (2019, June 3). A field-wide metasynthesis of pedagogical research in technical and professional communication. *Journal of Technical Writing and Communication.* https://doi.org/10.1177/0047281619853258

Mollick, E. R., & Mollick, L. (2022). *New Modes of Learning Enabled by AI Chatbots: Three Methods and Assignments* (SSRN Scholarly Paper No. 4300783). https://doi.org/10.2139/ssrn.4300783

Molohon, J., Reinsel, D., Strachan, T., & Stromback, C. (2019, December). *Non-traditional prosthetics* [Multimedia collection]. Fabric of Digital Life. https://fabricofdigitallife.com/Browse/objects/facets/collection:37

Museum and Library Services Act of 2010. (2010). Institute of Museum and Library Services. https://www.imls.gov/sites/default/files/mlsa_2010_2.pdf

Pedersen, I., & Aspevig, K. (2018). Being Jacob: Young children, automedial subjectivity, and child social media influencers. *M/C A Journal of Media and Culture, 21*(2). https://doi.org/10.5204/mcj.1352

Pybus, L. (2023, January 2). *Can ChatGPT write a proper usability test script?* https://uxdesign.cc/can-chatgpt-write-a-proper-usability-test-script-f8f6f89b5a2e

Ravichandran, S. (2022). *GPT-3: The AI writing tool every writer needs.* The Startup, Medium. https://medium.com/swlh/gpt-3-the-ai-writing-tool-every-writer-needs-43b22638a18d

Ryan, K., Loly, S., & De la Victoria, M. (2019, April). *What language sounds like: Wearable devices in translation communication* [Multimedia collection]. Fabric of Digital Life. https://fabricofdigitallife.com/Browse/objects/facets/collection:31

SciNote (2022). *Manuscript Writer by SciNote.* https://www.scinote.net/manuscript-writer/

Seeber, I., Bittner, E., Briggs, R. O., de Vreede, T., de Vreede, G. J., & Elkins, A., et al. (2020). Machines as teammates: A research agenda on AI in team collaboration. *Information and Management, 57*(2), Article 103174. https://doi.org/10.1016/j.im.2019.103174

Selber, S. (2004). *Multiliteracies for a digital age.* Southern Illinois University Press.

Sevilla, A., & Jennings, M. (2022, December 15). *Best AI writers of 2023.* https://www.techradar.com/best/ai-writer

Snyder, N., Burgan, P., & Wind, H. (2019, April). *Cultural reality—a VR experience* [Multimedia collection]. Fabric of Digital Life. https://fabricofdigitallife.com/Browse/objects/facets/collection:32

Sparrow, R., Dekoum, S., & Schaust, S. (2020, June). *Using technology to navigate foreign lands and cultures* [Multimedia collection]. Fabric of Digital Life. https://fabricofdigitallife.com/Browse/objects/facets/collection:54

Stambler, D. (2019, April). *Wearables and carryables for everyday communication: Past, present, future* [Multimedia collection]. Fabric of Digital Life. https://fabricofdigitallife.com/Browse/objects/facets/collection:35

Tham, J. (2019, January). *Emerging technologies for technical communication* [Multimedia collection]. Fabric of Digital Life. https://fabricofdigitallife.com/Browse/objects/facets/collection:29

Tham, J., Burnham, K., Hocutt, D. L., Ranade, N., Misak, J., Duin, A. H., Pedersen, P., & Campbell, J. L. (2021). Metaphors, mental models, and multiplicity: Understanding student perception of digital literacy. *Computers and Composition, 59*(March). https://www.sciencedirect.com/science/article/abs/pii/S8755461521000050

Tisza, N. (2023, January 9). *12 areas ChatGPT helps my work as a UX designer.* https://uxplanet.org/12-areas-chatgpt-helps-my-work-as-a-ux-designer-75d4dc6c99a2

Tuman, M. C. (1992). *Word perfect: Literacy in the computer age.* University of Pittsburgh Press.

Ulmer, G. L. (2003). *Internet invention: From literacy to electracy.* Pearson.

Vee, A. (2017). *Coding literacy: How computer programming is changing writing.* MIT Press.

Veeramoothoo, S. (2019, December). *Emerging technologies for business communication* [Multimedia collection]. Fabric of Digital Life. https://fabricofdigitallife.com/Browse/objects/facets/collection:42

Walton, R., Moore, K. R., & Jones, N. N. (2019). *Technical communication after the social justice turn: Building coalitions for action.* Taylor & Francis.

Waltzman, R. (8 August 2022). *Facebook misinformation is bad enough. The metaverse will be worse.* https://www.washingtonpost.com/opinions/2022/08/22/metaverse-political-misinformation-virtual-reality/

Winne, P. H., Vytasek, J. M., Patzak, A., Rakovic, M., Marzouk, Z., & Pakdaman-Savoji, A., et al. (2017). Designs for learning analytics to support information problem solving. In J. Buder, & F. W. Hesse (Eds.), *Informational environments: Effects of use, effective designs* (pp. 249–272). Springer. https://doi.org/10.1007/978-3-319-64274-1_11

7
PROFESSIONAL DIRECTION FOR HUMAN-AI INTERACTION

Overview

Chapter 7 begins with discussion of guidelines for professional practice surrounding human-AI interaction and includes a technical and professional communication (TPC) guide to human-AI interaction based on key studies in human-centered computing, science and engineering, and technical communication. Given the rapid increase in autonomous agents and human-AI interaction, the chapter's focus is on articulating autonomous agents and digital employees as augmentation technologies. After discussion of AI chatbots and conversational design, the chapter chronicles the development of digital employees across six corporations. Based on AI functional technologies, these digital employees are proposed to enhance meaningful, empathetic connections to the digital world, potentially resulting in more symbiotic relationships with users. TPCs must be positioned to intervene throughout the design, adoption, and adaptation of augmentation technologies. Audience is now an "augmented" audience; autonomous agents now function as independent team members; and content is produced through and with machine learning.

Key Questions

- What guidelines should influence professional practice surrounding human-AI interaction?
- How are digital employees (non–human autonomous agents) currently being designed and deployed?
- How are digital employees the next human-computer interaction interface?
- What impact do digital employees hold for TPC professional practice?

DOI: 10.4324/9781003288008-10

Chapter 7 Links

Throughout the chapter, we refer to articles, videos, and reports which can be found in a related chapter collection at Fabric of Digital Life called *Professional direction for human-AI interaction* (Duin & Pedersen, 2023). You can find a link to this collection at https://fabricofdigitallife.com.

Introduction

The year 2023 will be remembered as the year of the AI chatbot. With more than 100 million active users, ChatGPT has instigated a dramatic digital transformation (Hu, 2023). This chapter charts the emergence of digital employees, digital agents that will inform the next phase of the AI chatbot, construed as an augmentation technology.

In work foundational to our field, scholars Heidi A. McKee and James E. Porter (2017) pose one of their book's overarching questions, "what does it mean for professional communication that we're now talking to our computers?" (p. 21), pointing to how our relationships with technologies have changed. Later they query, "can AI bots be effective professional communicators?" (p. 135), asking whether artificial intelligence (AI) agents can stand in the place of human employees as communicators and eventually communicate for us. To position the autonomy of AI agents, McKee and Porter go one step further and ask "what about the smart agent who operates against the will of the human?" (p. 165), alluding to our ability to trust their actions. In this chapter, we take up these pivotal questions and chronicle the development of digital employees, tracing the turn toward their emergence in human and AI teams as augmentation technologies.

One of the key skills of a TPC is the ability to analyze and design communication for specific audiences/users. There are many ways to observe and interpret the needs of users including user experience (UX) tools and development of fictitious personas for representing target users. In earlier chapters we have discussed emotional enhancement. Emerging technologies like AI and body sensors are being used to track people's emotions, with automated emotion evaluation (AEE) deployed across robotics, marketing, education, and entertainment sectors (Dzedzickis et al., 2020), raising many issues for TPCs who work with data related to human emotions. Also as discussed earlier, questions about the reliability of facial recognition and other forms of sensors surround the use of these technologies.

Clearly there are both risks and opportunities with emotion AI (Purdy et al., 2019). Risks include bias built into systems affecting interpretations of facial expressions or body language, privacy concerns, and manipulation or exploitation of people's emotional responses for economic/social gain. Along with the risks come opportunities, where AI use for emotion recognition can track user engagement, and lessons can be adapted to personal interests. In healthcare, sensors can be used to detect warning signs and provide treatment options for people suffering with depression, anxiety, or behavioral disorders. For example, Limbic (https://limbic.ai) is a talk therapy platform (chatbot) that connects users to health resources, and WoebotHealth (https://woebothealth.com) is a popular app that uses psychological research and emotional design elements to help users with their mental health 24/7, without the need of making an appointment. Beginning with chatbots, these AI companions are rapidly being accepted by users and are becoming more mainstream, especially given their increasing therapeutic benefits. For example, the documentary *The Rise of A.I. Companions* (ColdFusion, 2022) traces the recent history and development of AI companions.

Each assistant's AI agency, defined as the capacity to act autonomously (act independently), to adapt (react and learn from changes in the environment), and to interact (to perceive and respond to other human and artificial agents), makes these useful (Dignum, 2019). In *Writing Futures* (2021), we explained how the popularity of virtual assistants has helped spawn the creation of "virtual humans" that are lifelike, virtual personas with nuanced facial, gestural, and spoken interaction. These screen-based, virtual humans mimic human physical reactions to be made to appear empathetic, unique, and mildly emotional.

To begin, we share guides to human–AI interaction.

Guides to Human-AI Interaction

Microsoft's Guidelines for Human-AI Interaction (2022) provide best practices for how an AI system should interact with people. Here one finds the HAX Workbook to drive team alignment across implementation; HAX design patterns to save time by describing how to apply established solutions during implementation; the HAX Playbook to identify and plan for common interaction issues; and useful examples in the HAX Design Library. The 18 guidelines, synthesized from 168 guidelines across two decades of research on how to make AI user-friendly, are grouped based on initial exposure to an AI system, interaction during use, support during system failure, and understanding over time:

- The initial phase
  - Make clear what the system can do.
  - Make clear how well the system can do what it can do.
- During interaction
  - Time services based on context.
  - Show contextually relevant information.
  - Match relevant social norms.
  - Mitigate social issues.

- When the system is wrong
  - Support efficient invocation.
  - Support efficient dismissal.
  - Support efficient correction.
  - Scope services when in doubt.
  - Make clear why the system did what it did.
- Over time
  - Remember recent interactions.
  - Learn from user behavior.
  - Update and adapt cautiously.
  - Encourage granular feedback.
  - Convey the consequences of user actions.
  - Provide global controls.
  - Notify users about changes.

(Microsoft, 2022)

Apple, Google, and Microsoft have each released AI interaction guidelines. Austin Wright and colleagues from the Georgia Institute of Technology and the University of Konstanz (2020) completed a comparative analysis of the three sets of guidelines along with an inclusive taxonomy of how all the guidelines fit together. Instead of organizing guidelines around the process for a user (Microsoft's approach), Google focuses on "distinct concepts that a *developer* has to continuously keep in mind. These are: User Needs + Defining Success, Data Collection + Evaluation, Mental Models, Explainability + Trust, Feedback + Control, and Errors + Graceful Failure" (p. 2), resulting in 113 guidelines. In contrast, Apple focuses on *practitioner knowledge*, "foregoing references or data, and thus... seemingly based entirely upon design principles within the Apple organization" (p. 2), resulting in 59 guidelines. Developed by Wright et al., Figure 7.1 illustrates the relative emphasis of human-AI interaction guidelines across the three companies, with a "dot" indicating no emphasis for a guideline

FIGURE 7.1 Wright et al.'s (2020) summary of Apple, Google, and Microsoft human–AI interaction guidelines.

Source: Used with permission of Austin P. Wright.

category. Overall, Google emphasizes "model considerations for training data and processes" (p. 4); Microsoft emphasizes mental models; and Apple emphasizes "smooth user experiences such as Error Prevention, Calibration, Confidence, and Multiple Options" (p. 4). This figure can be used to launch TPC discussion, design, adoption, and adaptation of augmentation technologies.

Described as a "loose collection of things that may be important to keep in mind," Wright et al. also promote a community-generated set of guidelines that can be found at https://ai-open-guidelines.readthedocs.io/en/latest/.

These sets of guidelines understandably focus on business product development along with the need for diverse perspectives, transparency and communication, and human control. As none of these sets draws directly from technical communication and related disciplines, we created the following TPC guide to AI-human interactions based on mapping the work of key scholars across three related disciplines: Duri Long and Brian Magerko (2020) in human-centered computing, Jason Chew Kit Tham (2018) in technical communication, and Saleema Amershi et al. (2019) from science and engineering (see Table 7.1). In each case, the scholar's or scholars' goal was to create a resource for practitioners and scholars working on the design, adoption, and adaptation of immersive/AI technologies.

We discussed Long and Magerko's set of competencies and design considerations in detail in Chapter 4; again, their conceptual framework is derived from a synthesis of interdisciplinary literature. Tham's study of "Interactivity in an Age of Immersive Media" is also based on a review of key literature from which he proposes seven constructs or dimensions of interactivity for the design of immersive media, mapping these to the interactive design of wearables and IoT products. He writes that:

> the new dimensions of interactivity in immersive media environments suggest that technical communicators and designers must pay attention to 1) the seamless connection between the user and the systems, data, and actions; 2) user control and customization; 3) proactive contextual assistance from smart technologies; and 4) device sensibility and sensory simulations.
>
> *(p. 46)*

Amershi, in collaboration with 12 colleagues in "Guidelines for Human-AI Interaction," determined their design principles from a study of 49 design practitioners who tested 20 "popular AI-infused" products. Including collaborators from Microsoft, Amershi et al. group their resulting 18 design guidelines based on a user's initial exposure to the AI, initial interaction, understanding of what went wrong, and use over time.

The guiding principles discussed in Chapter 5 parallel these guidelines; however, Chapter 5's discussion focuses more in depth on the socio-ethical consequences inherent in privacy, accountability, safety and security (including

TABLE 7.1 AI interaction guidelines for TPCs based on themes from Long and Magerko (2020), Tham (2018), and Amershi et al. (2019)

User control / Autonomy	Explainability	Data Considerations
• Allow user customization. • Create devices with multiple forms of input and feedback. • Give users autonomy and agency.	• Distinguish general and narrow AI, making clear what the system does. • Convey consequences of user actions.	• Understand basic data literacy concepts, recognizing that computers learn from data. • Investigate who collected and created the dataset, along with its limitations.
Computational Thinking	**Appropriate Use Contexts**	**Emotional Design**
• Understand steps involved in machine learning, including the human role in fine-tuning AI systems. • Focus on visual / auditory elements and fill-in-the-blank coding.	• Consider learners' identities, values, age, and backgrounds for interest in AI. • Provide scaffolding and support; match relevant social and cultural norms.	• Leverage learners' interests when designing AI literacy interventions. • Focus on entertainment, encouraging participation.
Embodied Contexts	**Action and Reaction**	**Ethics**
• Involve embodied simulations of algorithms, hands-on experimentation with AI technology. • Work to understand affordances and limitations of various modalities.	• Allow users to inspect system components, explaining few at once. • Include contextually relevant information. Make it easy to dismiss undesired AI systems.	• Identify key issues: privacy, employment, misinformation, diversity, bias, transparency, and accountability. • Question AI intelligence and trustworthiness.

harassment and enhancement vulnerabilities), transparency and explainability, and fairness and non-discrimination. We encourage practitioners to use Table 7.1 together with Chapter 5's principles as a launching pad for work on human-AI interaction. While attention to user control and appropriate use contexts are regularly practiced in TPC, we suggest that practitioners expand on ways to help users

• Distinguish between general and narrow AI (explainability);
• Recognize how data is collected and how computers are learning from data (data considerations)—including understanding of machine learning and the human role in fine-tuning systems (computational thinking);

- Experiment with AI technology through embodied simulations of algorithms (embodied contexts);
- Inspect system components, assuring the ability to dismiss undesired AI systems (action and reaction); and most important,
- Question AI intelligence and trustworthiness (ethics).

For example, recall the discussion in Chapter 2 of the Wearable Reasoner, an AI assistant device for enhancing judgment and decision making in real time, encouraging people to be more skeptical and to seek and demand evidence before jumping to conclusions (Danry et al., 2020, 2021). This device pairs smart glasses (Bose Frames) with open-ear speakers that enable private audio feedback while not blocking the ear canal. The device has a microphone that detects utterance input and it can be connected via Bluetooth to a smartphone for additional processing through a mobile application. This wearable, audio-based system examines logical structures of an argument and offers real time evaluative and analytic feedback to the wearer through its use of "an algorithm capable of delivering relevant feedback on argumentative structures" (p. 4).

Next, consider the above AI interaction principles in relation to the Wearable Reasoner device. By doing so, we can work to better understand the device's use of general and/or narrow AI (recall discussion of this in Chapter 4), how the data is being collected via utterance input and how the device learns from the data, and the specific role of machine learning together with the human role in fine-tuning the system. We can ask how users might inspect Wearable Reasoner components to determine and possibly dismiss undesired AI systems. And most importantly, we can help users to identify both correct information as well as misinformation as a means to build trust in the device. Correctly distinguishing correct information from misinformation depends on overall understanding of the social and cultural context surrounding deployment of the system.

Human-AI Teaming

In November 2021, media outlets circulated news about a coming era of the Metaverse that will change work, school, and commerce on a mass scale. We discuss this development in Chapter 3 and the means through which this hype serves to justify investment in these technologies. One *Forbes* article predicts, "The biggest change is that social media is going to become a virtual personification of real life — think shopping malls, concert halls, and 'virtual humans' that look exactly like us" (Brandon, 2021). The repetitive claim is that virtual experiences will become "real life" and that personified agents will "look like us." Humans will have avatars that look like humans, and AI agents will have virtual bodies that also look like humans. Put another way, the intent is to evolve virtual workers not as cartoon-like characters, but rather as photorealistic entities. The Director of Strategy at Star Labs explains, "Neons are computer-generated virtual beings, artificial humans modeled after real people" (Lian, 2020).

However, this transformation is more than a drive for greater efficiency and automation; it also involves transforming notions of trust. In our previous book, we ask "What happens when we cannot tell the difference between our human co-workers and our AI agents?" (Duin & Pedersen, 2021b, p. 88). We make the point that "the goal to design hyperreal 'physical' qualities for virtual humans responds to the need for legitimacy (ethos) for virtual assistants to communicate in real life scenarios, such as helping patients navigate a hospital" (p. 88). Our exemplar is Rachel by Quantum Capture, a current virtual human technology that serves as a hospitality host or concierge (discussed in more detail later in this chapter). The online marketing materials explain: "Humanizing AI increases its effectiveness and drives adoption. 55% of human communication is non-verbal."

In *Writing Futures*, we emphasized the exigence for planning for futures that include AI writing assistants and AI collaborators, with specific attention toward the evolution of technology "from assistant to collaborator, from machine autonomy to human-machine cooperation and collaboration, from assemblages to collaboratives, from assistantship to synergy" (p. 10). In the next section, we trace the emergence of human-AI teaming scenarios framed to augment human tasks through the deployment of digital employees. While this type of technology uses differing terminology—"artificial humans," "digital people," "digital humans," "virtual humans," and "digital employees"—we argue that these categories describe a progressed version of more common digital assistants, e.g., Siri, Alexa, and Google Assistant. For the Gartner organization's discussion of 2021 hype cycle themes, Kasey Panetta (2021) writes:

> Consider digital humans, which are digital-twin representations of people. This technology presents an opportunity for licensed personas that enable new revenue streams. They can appear as avatars, humanoid robots or conversational user interfaces, like chatbots or smart speakers. These interactive, AI-driven representations seem human and behave in 'humanlike' ways supported by a range of technologies including conversational UI, CGI and 3D real-time autonomous animation.

We use the umbrella term "digital employee" to define this emergent technology, not to promote its use but to make transparent the goal for these companies.

Digital employees mimic human behaviors on the one hand, and they also automate work tasks on the other with extensive AI capabilities that humans would not be able to perform. In line with the goals for this book, we argue that through proxy agency, digital employees augment human cognitive capabilities using this form of personified AI agent. Discussed in Chapter 3, agents "exercise proxy agency through a technologically mediated mode referring to the entanglement of human and non-human agencies," which "implies an obligated relationship" (Neff & Nagy, 2019, p. 102).

Digital employees are designed to collaborate with human employees but are also sold by third-party companies for enterprise services. They are designed to emote through verbal, facial, and gestural responses, and many are equipped to sense or detect emotion. From a socio-ethical standpoint, society needs to understand their role. They *do the work* of humans; they collaborate with us; they learn from and with us.

Instructors have long worked to introduce technology in TPC classes, with increasing focus on markup languages and content management, and now, writing for machine translation. We have long taught about audience—addressed, invoked, engaged, involved, immersed—with the primary assumption always being that of a human audience. Breeanne Matheson Martin (2016) stressed that:

> teaching students to carefully consider the human audiences that will encounter their content is not an adequate strategy to enable students to do technically sound, ethical work. Instead, students must also be equipped to evaluate the machine audiences that may read, interpret, translate, or evaluate the texts they create as well as the human ones
>
> *(p. 1)*

Articles by John Gallagher (2017, 2020) also provide direction for teaching students about the role of algorithms as an audience.

Likewise, professionals work to incorporate and use emerging technologies to better understand audiences (data analytics) and create greater efficiencies and effectiveness of documentation (component content management). Given the exponential rise of AI and augmentation technologies, we turn next to providing professionals with further direction for interaction with autonomous agents.

Autonomous Agents as More than Tools

To begin, TPCs must see autonomous agents as more than tools. As noted in Chapter 1, John Novak and colleagues (2016) ask, "When do electronic tools cease to be 'simply' tools, and become meaningfully part of ourselves?" From this pivotal question, they add "When might we think of these tools as augmenting ourselves, rather than simply amplifying our capabilities?" Here we push this even further, asking, "When do autonomous agents decide everything and act autonomously?"

As part of conference presentations on human-autonomy teaming (HAT) (Duin & Pedersen, 2021a; Pedersen & Duin, 2022), we have found Thomas O'Neill et al.'s (2022) synthesis of 76 empirical studies on HAT to be particularly helpful. We encourage TPCs to discuss the following ten levels as a way to identify low to high automation and agent autonomy levels:

10 The computer decides everything and acts autonomously, ignoring the human.
9   The computer informs the human only if it, the computer, decides to.
8   The computer informs the human only if asked, or
7   The computer executes automatically, then necessarily informs the human, and
6   The computer allows the human a restricted time to veto before automatic execution, or
5   The computer executes that suggestion if the human approves, or
4   The computer suggests one alternative, or
3   The computer narrows the selection down to a few, or
2   The computer offers a complete set of decision/action alternatives, or
1   The computer offers no assistance; the human must take all decisions and actions. (O'Neill et al., 2022, p. 909)

O'Neill et al. define an autonomous agent within this HAT context as:

> a computer-based entity that is recognized as occupying a distinct role on the team. The autonomous agent is more likely to be recognized as a team member if humans and autonomous agents are interdependent and the autonomous agents are perceived as agentic.
>
> *(p. 911)*

They define human-autonomy teaming as:

> interdependence in activity and outcomes involving one or more humans and one or more autonomous agents, wherein each human and autonomous agent is recognized as a unique team member occupying a distinct role on the team, and in which the members strive to achieve a common goal as a collective. The 'autonomy' aspect of human–autonomy teaming refers to the autonomous agent.
>
> *(p. 911)*

Their findings indicate that higher levels of agent autonomy as well as higher levels of interdependence provide for better outcomes. Not surprisingly, they find that:

> communication among autonomous agents and humans tends to be different than communication among humans... autonomous agents need to better anticipate human information requirements and communicate

the right information in a timely fashion (pushing behaviors), rather than not anticipating and requiring a human inquiry for the information and a lag time (pulling behaviors).

(p. 927)

In short, the autonomous technology searches for data, transforms the information, and makes subsequent decisions or processes that result in something being done for a human that the human then no longer needs to do. Each increasing level represents increasing automation; the human role increasingly becomes that of a supervisor. At the highest levels of autonomy, we may still design the agents, but they no longer need us to function.

Ewart deVisser et al. (2018) state that autonomy is "a capability that does not need a human to function," and that humanness design is "a capability designed to connect with a human." They plot existing and fictional agents according to the high and low degrees of autonomy and humanness design; these result in the following four quadrants:

- *Tools* (low autonomy and low humanness design) include Alexa, Siri, DeepBlue or tools designed to play strategy games, and self-driving cars;
- *Companions* (low autonomy and high humanness design) include Huggable Robots, companion robots, Samantha (*Her*), and David (*AI*);
- *Superintelligent entities* (high autonomy and low humanness design) include Skynet (*Terminator*), H.A.L. (*2001: A Space Odyssey*), and Replicants (*Blade Runner*); and
- *Spiritual machines* (high autonomy and high humanness) include Data (*Star Trek*), Simulated Prophet, and Oracle (p. 1411).

deVisser et al. stress the need to increase humanness "if the design requires a connection and communication with a human user," noting that increased complexity and potential mismatch between user needs and AI capabilities results in "humanness" as "a required interface feature." Plotting these on the O'Neill et al. scale places "tools" at the lowest levels, "companions" at mid levels, and "super intelligent entities" and "spiritual machines" at the highest levels of autonomy.

The next section of this chapter traces this current turn to human–autonomy teaming. For manufacturing CEO and co-founder of Veo Robotics Patrick Sobalvarro (2020) chronicles the initial turn, stating that "The best choice is to combine the strength, precision, and speed of industrial robots with the ingenuity, judgment, and dexterity of human workers." He quotes recent research on human–autonomous robot teaming that indicates a 95% increase in productivity "when people work collaboratively with a human-aware robot compared to when working in all-human teams." Likewise, TPCs have well adapted to coexisting with multiple technologies and to increased human-technology combinations. For example, Alex Masycheff (2018), writing about how to consider dividing responsibilities with robots in technical communication, suggests use of natural language processing (NLP) technologies to create knowledge maps for guiding users through mounds of information; automated metadata tagging to improve the quality of technical documentation; and dynamic documentation assembly. Our next step is to adopt and adapt with digital employees.

Tracing the Turn to Human-Autonomy Teaming

In Chapter 4, we noted Tori Homann's (2022) identification of automation and augmentation trends:

> Even though the thought of robots and AI taking over some tasks of the workforce can be scary, there is also a lot of growth potential.... By removing the burden of repetitive or low-level problem-solving tasks, workers are freed up to contribute to value-adding tasks. So when a worker performs in tandem with automated technology, their job is augmented so that they can accomplish even more work in less time with fewer mistakes.

The Deloitte Corporation calls this worker plus automated technology a superjob; North American Signs calls it a superteam (Mallon et al., 2020).

Chatbots

Also in Chapter 4, we wrote about how TPCs in user experience and support specialist roles will find their positions changing amid algorithmic AI technologies. We discussed the need to position chatbots between front-line technicians and self-help or self-service, as the chatbot provides an interface where users can interact with the information in a more natural way. The user receives direction, and the chatbot learns from the user. An additional example comes from Hans van Dam (2018), founder of the Conversation Design Institute, who illustrates conversation design for chatbots (see Figure 7.2) as a way to emphasize that "the Conversation Designer has to understand both the human and the artificial brain, and he has to use copywriting techniques to make sure that both brains understand each other."

Companies like Contentstack and Adobe (Adobe Enterprise Content Team, 2020) provide help for businesses transitioning to AI-enhanced content management. Because AI can analyze large amounts of data and automate tasks, it is increasingly employed as part of content analysis, management, and creation. AI can be used to tag text and images in documents, analyze tone or sentiment in text, improve personalization by pushing content to users based on their search history, make predictions, and create content.

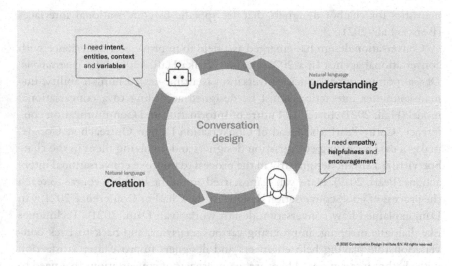

FIGURE 7.2   Conversation design for chatbots.
Source: Used with permission of Hans van Dam.

In the future, AI capabilities will improve as connections between devices increase (Internet of Things, IoT). This improvement in connectivity will be seen in content management systems, platforms, and tools as well, leading to many benefits for consumers, businesses, and organizations. Content managers will have access to more data and will have more efficient ways of accessing and distributing content (Gauss, 2021). As the use of AI-based chatbots continues to expand, customers are able to get help from non-human agents 24/7.

Chatbots may replace many customer service representatives, presenting an important role for TPCs in the design phase of a chatbot. In a recent webinar documented by Barb Zinck (2022), she discusses Ann Rockley's direction for managing content for chatbots and AI, pointing out that interactions with chatbots involve three main elements that coincide with the skills of TPCs. A human–chatbot interaction always involves context (a reason for the interaction), intent (a purpose or goal), and entity (how the content is delivered). An entity can be an information component or larger piece of tagged content that can be used by an AI system to streamline the process of moving the right content to the user interacting with the chatbot. Structured content, or combining AI with a component content management system (CCMS), can make it easier for TPCs to update and reuse information that will be shared in a chatbot-user interaction.

Industries will continue to adopt chatbots as a way to meet customer needs (Yuen, 2022). As technology for chatbots and virtual assistants improves, users are more likely to view interactions with them favorably (Lohr, 2022). To measure the quality of interactions with chatbots, researchers are developing heuristics for chatbot designers that are specific to conversational interfaces (Borsci et al., 2021).

Conversation design has emerged as a field to improve user experience with conversational agents. In a 2021 webinar, Erika Hall, author of *Conversational Design*, points out that since conversation is a time-tested human ability, human-computer interactions must be designed according to a conversational model (Hall, 2021). In a 2019 Future of Information and Communication conference, Cathy Pearl, the Head of Conversation Design Outreach at Google, makes a case for hiring conversation designers and involving them in the chatbot/virtual assistant design early in the process to improve conversational interactions (Pearl, 2019). Hans van Dam started an online learning course to teach the process of good conversation design. At the Chatbot Conference 2021, van Dam explained how conversation design works (van Dam, 2021). Techniques like dialogue mapping, improvising various scenarios, and back-to-back conversation role playing help engineers and designers improve user satisfaction with chatbot interactions. There are many aspects of conversations that need to be taken into account in the design process like personality, error management, turn taking, initiating a conversation, managing the user's expectations, and meeting the goals of the conversation in a timely manner.

Conversation designers come from many different fields such as copywriting, psychology, UX writing, linguistics, communication, and audio design. This multidisciplinary work combines human communication and technology in ways that would be familiar to TPCs who are used to thinking about rhetoric and the design of human-computer interactions. Imagining the future of voice user interfaces (VUI), Fountain Institute co-founder Jeffery Humble sees many benefits to computer technologies becoming "a bit more human." At a 2018 conference, Humble imagines VUI devices and tools that are more responsive to our moods, needs, and verbal requests (Humble, 2018). This idea that conversation is a good model for human-computer interactions is foundational in technical and professional communication (e.g., see Redish, 2012; Turkle, 2016).

In short, TPCs as conversation designers need to make chatbots more helpful, natural, and persuasive; TPCs as co-designers of and co-workers with digital employees also must pursue ethical ends for designing helpful, natural, persuasive, and trustworthy digital employees. In contrast to chatbots, digital employees must be taught to have a conversation about the information they are learning (VirtualHumans.io); they must be able to talk "on demand" with another person. In so doing, they emerge as digital employees.

TPC work is somewhat related to the study of human-AI friendships resulting from interaction with social chatbots. Petter Brandtzaeg et al. (2022) recently interviewed 19 persons who have human-AI friendships with the social chatbot Replica, finding the artificial nature of the chatbot to alter the notion of friendship in multiple ways. Likewise, the increased use of chatbots across

TPC is bound to alter the notion of documentation development and usability in multiple ways. Brandtzaeg et al. note that these social chatbot systems are "capable of having social and empathetic conversations with users" (p. 404), resulting in a more intimate relationship with technology. Again, instructional chatbot systems designed for engagement with technology may well result in a stronger connection with the technology and tasks being completed.

Emergence of AI Digital Employees

This section traces the emergence of AI digital employees, identified as artificial humans, digital humans, virtual humans, digital employees, and soul machines. We begin with NEONs (2022, https://neonlife.ai), first introduced by Samsung at the Consumer Electronics Show (CES) in Las Vegas 2020 and considered the world's first **artificial human**. Described as "a computationally created virtual being that looks and behaves like us," in corporate videos, NEONs are AI-powered avatars with a whole human-like moving body representation. These digital avatars are powered by artificial intelligence and thus able to appear to sympathize and converse like humans. In Table 7.2, we share its corporate terminology and embodied representation, its persuasive claims in videos, how its AI functions, and the work domains for which NEONs are being positioned. In an earlier rendition of this site, Samsung Star Labs

TABLE 7.2 Qualities of Samsung Star Labs' NEONs

NEON Corporate terminology	Extent of embodied representation in corporate videos	Persuasive claims about digital employees	Discourse on AI functional technologies	Work domain— examples
The world's first **artificial human** "NEON" AI-powered avatar	Whole human-like moving body	"They're virtual and they behave just like you and me. They can smile, they can talk, they can laugh. They have expressions and the idea was to make technology more like us to make machines more like us" (Neon, 2020).	Natural language processing Behavioral neural networks Core R3 for reality, real time, responsiveness	Humanoid chatbot Customer service, makeup/ beauty consultants, weather reporters, yoga instructors, travel assistants, financial advisers, translators

claimed that NEONs can "converse and sympathize like real human beings" as a result of their use of NLP, behavioral neural networks, and additional proprietary technology called Core R3 for reality, real time, and responsiveness. As essentially humanoid chatbots, the current site positions NEONs for service and adviser roles. However, although being considered the world's first artificial human, and despite its impact on development of autonomous agents, this subsidiary of Samsung provides less detail about its most recent work. In a case like this, Fabric serves as a critical resource that continues to include original artifacts related to NEONS (Star Labs, 2022).

In order to humanize AI while increasing effectiveness and adoption, Quantum Capture (https://www.quantumcapture.com), slated to "drive meaningful customer connections with intelligent digital humans," uses the term **digital humans** (see Table 7.3). Figure 7.3 presents the digital human, Aiko. Quantum Capture emphasizes the role that body language, eye contact, posture, and emotions play in how humans interact and absorb information, writing that "digital humans elevate todays chatbots and are quickly becoming the ultimate user interface." While the "chatbot foundation" provides a "highly scalable, multilingual, 24/7 availability, consistent & accurate, professional & polite" foundation, they promote digital humans with the ability to deliver "non verbal communication, increased emotional engagement, physical brand identity, and product visualization." For example, their digital ambassadors are slated as the "Zeitgeist" to "capture the imaginations of the public, deliver presentations on-brand and on-message and interact with people in ways never before thought possible." These digital humans represent whole human-like moving bodies, designed using NLP, Unreal Engine, and JavaScript for two-way communication. The site states that a digital human "is designed specifically to pair high quality digital human characters together with advanced 3rd

TABLE 7.3 Qualities of Quantum Capture's digital humans

Quantum Capture Corporate terminology	Extent of embodied representation in corporate videos	Persuasive claims about digital employees	Discourse on AI functional technologies	Work domain—examples
Digital human	Whole human-like moving body Digital avatars, chatbot, text to speech, SDK, web client front-end (user interface)	"We enable you to give a body to an AI, like a Watson or Google Home or Alexa, with a body." (Quantum Capture, 2018).	Natural language processing (IBM Watson)	Customer service, retail, hospitality, healthcare, banking, entertainment Product specialists, brand & marketing ambassadors

FIGURE 7.3   Aiko, an example of a digital human.
Source: Used with permission of Quantum Capture. https://www.quantumcapture.com.

party natural language and AI processing services… [so] that no special gaming hardware is required" for its use (Quantum Capture, 2022a). Fabric serves as a critical resource for original artifacts related to these digital humans (Quantum Capture, 2022b).

The New Zealand firm UNeeQ (https://digitalhumans.com) also pro-motes **digital humans** (see Table 7.4) as a means to "revolutionize customer experiences with scalable human connections," defining digital humans as "AI-powered customer experience ambassadors that recreate human interaction at infinite scale" (UNeeQ, 2022a). UNeeQ initially launched Digital Einstein, a virtual double and AI companion that users can speak with about science. For this next project, users can immediately meet and chat with Sophie (https://iamsophie.io, see Figure 7.4). Similar to Quantum Capture, UNccQ has lent Sophie's likeness to BMW, Deloitte, IBM, and Deutsche Telekom as a vir-tual brand ambassador. Launching her own non-fungible asset (NFT) project, Sophie wants visitors to buy her NFT art (5,555 different, short videos), "all of which the AI created after 'training' as an artist–of the bot dressed in one of 12 outfits" (Harley-McKeown, 2022). Users have to follow her on Twitter and Discord and gain access to NFTs as a means to meet later with Sophie in her metaverse art studio. Here designers have access to step-by-step instructions for a basic open source example of a digital human interface (https://gitlab.com/uneeq-oss/examples). For persuasive claims, UNeeQ emphasizes that "Digital humans bring meaningful connection to the digital world, where empathy and compassion have disappeared from customer interactions." Fabric serves as a critical resource for original artifacts related to UNeeQ (2022b).

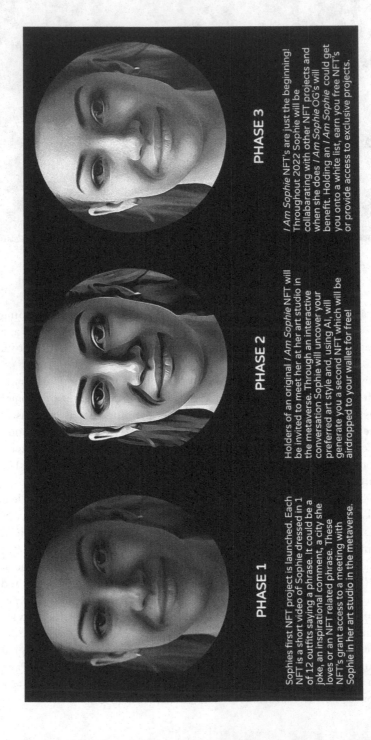

PHASE 1

Sophies first NFT project is launched. Each NFT is a short video of Sophie dressed in 1 of 12 outfits saying a phrase. It could be a joke, an inspirational comment, a city she loves or an NFT related phrase. These NFT's grant access to a meeting with Sophie in her art studio in the metaverse.

PHASE 2

Holders of an original *I Am Sophie* NFT will be invited to meet her at her art studio in the metaverse. Through an interactive conversation Sophie will uncover your preferred art style and, using AI, will generate you a second NFT which will be airdropped to your wallet for free!

PHASE 3

I Am Sophie NFT's are just the beginning! Throughout 2022 Sophie will be collaborating with other NFT projects and when she does *I Am Sophie* OG's will benefit. Holding an *I Am Sophie* could get you onto a white list, earn you free NFT's or provide access to exclusive projects.

FIGURE 7.4 Phases of Sophie's non-fungible asset project.

Source: Used with permission of I Am Sophie, UneeQ Ltd., https://iamsophie.io.

TABLE 7.4 Qualities of UNeeQ's Sophie, a digital human-AI-powered virtual influencer

UNeeQ Corporate terminology	Extent of embodied representation in corporate videos	Persuasive claims about digital employees	Discourse on AI functional technologies	Work domain—examples
Digital human AI companion, Digital Einstein AI-powered virtual influencer, Sophie	Whole human-like moving body, full body interaction	"I am the world's most advanced Digital Human because I am animated in real time using Artificial Intelligence" I Am Sophie. (2022).	Open AI, GPT-3 deep learning neural network architecture	Company focus on solving problems for brands, healthcare, education Brand & marketing ambassador Fashion model, NFT dealer

While digital humans are marketed for service sectors, Virtro Technologies Inc.'s (https://www.virtro.ca, see Table 7.5) mission is "to accelerate learning by leveraging innovative AI and VR technology." Those interacting with **virtual humans** are to think of them "as a friend that you can speak with, learn from, or even have as a guide in a learning situation." Also built using NLP and machine learning, these virtual humans are "programmed and trained to deliver conversations around specific topics, making them particularly useful in training simulations." The Virtro site indicates nine use cases for virtual humans (https://www.virtro.ca/virtualhumans); in addition to customer service, these include simulations for job interviews, healthcare practice, language practice, press and media work, performance reviews, custom training, and trades and manufacturing. An additional use case is titled "digital twin technology" (Virtro Technology Inc, 2022a). For delivery, Virtro uses VR or WebXR training applications. Persuasive claims include their 24/7 availability, lower cost, training consistency, and use of smart analytics for personalized help and support. See Fabric for original artifacts (Virtro Technology Inc, 2022b).

As noted earlier, we refer to the above digital entities as **digital employees**. Amelia (https://amelia.ai/conversational-ai/) (Amelia US LLC, 2022a), to our knowledge, is one of the first "conversational AI" entities marketed as a digital employee and conversational AI solution (see Figure 6.2 and Table 7.6). Unique here is Amelia's Digital Employee Builder that uses AI to build AI. The persuasive claim is that this "'Democratizes Conversational AI' by allowing human employees to create and customize their agent using natural language, without any specialized coding skills." The Amelia videos promote this ability to create digital employees through statements including the following:

TABLE 7.5 Qualities of Virtro Technologies Inc.'s virtual humans

Virtro Corporate terminology	Extent of embodied representation in corporate videos	Persuasive claims about digital employees	Discourse on AI functional technologies	Work domain—examples
Virtual humans	Whole human-like moving body	"Virtual Humans are patient, noon-judgmental, and are always available" (Virtro VR and Virtual Human training simulations, 2021, April 26).	Natural language processing Machine learning	Training simulations; practice partners; Teaching career leadership skills, language learning tutor

TABLE 7.6 Qualities of Amelia (US LLC)

Amelia Corporate terminology	Extent of embodied representation in corporate videos	Persuasive claims about digital employees	Discourse on AI functional technologies	Work domain—examples
Digital employee "Amelia"	Human-like face and shoulders	"Democratizing Conversational AI Create your Digital Employee, using an interactive interface, for a specific role. The Digital Employee Builder is built with the business-savvy in mind, so the process requires no software coding experience." (Amelia.AI, 2020).	Natural language processing Emotion detection	IT Office Help desk System upgrades Retail, Banking, customer service

- "The most human AI is now also the easiest to set up and use"
- "The new tool uses Amelia to build Amelia"
- "Speak or write in natural language and Amelia will build your digital employee for you"
- "Give it a name, select an area of specialization, add skill buttons"
- "Use the paraphrase button to generate phrases for you"
- "You are using AI to build AI, so you do not need to know the proper protocols to follow"

- "Amelia interprets entire statements and creates forks for you"
- "Amelia learns the content of the answers, more than just the answers themselves"
- "In just a few minutes, you were able to create a digital employee"

Artificial Intelligence for IT Operations (AIOps) is an "ITIL-compliant, end-to-end intelligent automation and management solution for IT operations." Those interacting with Amelia can perform tasks and resolve problems which she in turn either completes on her own or with assistance from human engineers. See Fabric for a collection of Amelia artifacts (Amelia US LLC, 2022b).

Using the term **digital people**, Soul Machines Limited (2021, https://www.soulmachines.com) allows visitors to its site to immediately experience talking with digital people, in this case with Viola (Soul Machines Limited, 2022a) (see Figure 7.5 and Table 7.7). Its use cases span a broad spectrum that includes customer service, healthcare, financial services, brand ambassadors, entertainment, education, real estate, policing, and telecommunications/call centers. In the education sector, soul machines are positioned to answer financial, course, career, and guidance questions; to serve as administrative or virtual teaching assistants; and to speak in 12 languages:

> Digital People can hear, see, understand, think and relate to customers.... In education, their empathic human behavior transforms impersonal online interactions into meaningful visual connections making the students feel supported and mentored through their academic careers.

Say hello to Digital People

Your journey to the metaverse starts here.

☐ I understand how to get the best experience talking to Viola and give consent to use my camera and mic

TALK TO VIOLA

FIGURE 7.5 Soul Machines' Digital People.
Source: Used with permission of Lisa Hendersen, Soul Machines, https://www.soulmachines.com.5.

TABLE 7.7 Qualities of Soul Machines

Soul Machines Corporate terminology	Extent of embodied representation in corporate videos	Persuasive claims about digital employees	Discourse on AI functional technologies	Work domain—examples
Digital People Digital employees Humanized AI	Human-like face and shoulders	"We are making the leap from facial gestures to fully bodied gestural performance. I will be able to use my whole body to communicate from head to toe." Soul Machines (2021, May 26).	Natural language processing (IBM Watson) Emotion detection	Customer service, healthcare, financial services, brand ambassadors, entertainment, education, real estate, policing, telecommunications/ call centers

Greg Cross, CEO of Soul Machines, shared the following with us:

> Using ground-breaking AI and natural language processing, Soul Machines is creating Digital People like Viola, to reimagine communication and engagement. Digital People present uncapped potential for brands to interact with customer bases, fans and more through empathetic and personalized experiences that are tailored to an individual's culture, language, and location. We are in a transformational era and when designed and implemented ethically, AI will facilitate authentic connections that enrich our everyday lives. (Conveyed by Nichole Rondon, personal communication, Soul Machines' Marketing/PR firm.
>
> *(DKC News, July 26, 2022)*

See Fabric for examples of Soul Machines artifacts (Soul Machines Limited, 2022b).

These six examples illustrate augmentation technologies designed for users' simultaneous cognitive, sensory, and emotional enhancement through two-way open communication. Based on AI functional technologies, these digital employees provide customer service across multiple sectors, in each case offering increased connections via "ideal proactive companions." Digital employees are tireless; moreover, they are proposed to "bring meaningful connection to the digital world, where empathy and compassion have disappeared from customer interactions" (https://digitalhumans.com). As compared to previous technologies, they bring the greatest autonomy and agency to date to the agent.

Our study of digital employees did not include a gender analysis. However, we note that four of the six examples feature digital employees exhibiting visual traits classified as those associated with women. We note that this is an area of the research that could benefit from further study with more attention to implications across the full gender spectrum. Further, we note the use of racialized representations in the advertising and actual interface of digital employees. Contextualization of race should be interpreted with input from communities, citizens, and professionals inculcated by the outcomes of these depictions, as an act of coalition-building. We point TPCs to experts in the field. Willa-jeanne F. McLean, a distinguished professor of law and an internationally recognized expert on the interface of intellectual property, writes about bodies of color and the issues of cultural appropriation and commodification (2021). Lisa Nakamura, a professor of visual culture and media, has contributed an extensive list of works in the area, including the pioneering book on the topic, *Digitizing Race* (2007). Black in AI is a non-profit organization helping to bring Black talent to the development of AI (https://blackinai.github.io). In short, ethical principles *Inclusiveness in Design* and *Inclusiveness in Impact* (see Chapter 5) require consideration in order to avoid discriminatory implications surrounding all aspects of digital employees.

Symbiotic Agency

As shared in Chapter 3, prominent posthuman theorist N. Katherine Hayles (2002) states that "agency still exists, but for the posthuman it becomes a distributed function" (p. 319). She clarifies our lived experience as follows:

> Living in a technologically engineered and information-rich environment brings with it associated shifts in habits, postures, enactments, perceptions—in short, changes in the experiences that constitute the dynamic lifeworld we inhabit as embodied creatures.
>
> *(p. 299)*

Reaffirming her position in a later book (2017), she explains post-digital human and non-human collaboration in terms of "complex human technical assemblages in which cognition and decision making powers are distributed throughout the system. I call the latter cognitive assemblages" (p. 4). We add the term "cognitive enhancement" to our evolving understanding and use of augmentation technologies.

As discussed earlier, Gina Neff and Peter Nagy (2019) speak of the "agentic" behaviors that come as a result of machine learning, artificial intelligence, and chatbots. Now, with digital employees, ever more complex, intimate relationships expand between users and technologies. "Agency has long connoted a distinctive human ability and something that by definition only humans possess" (pp. 97–98). They note how Bruno Latour (1999) and François Cooren (2010) theorized about the tensions between human and non-human agents, working to "redefine agency in terms of how and when people interact with complex technological systems" (p. 98). They draw on Albert Bandura's (2001) social-cognitive psychological theory of human agency to understand this new role of users in "shaping and determining technical systems," using the term "symbiosis" for the increasingly interdependent relationships with technologies such as digital employees. Neff and Nagy's pragmatic working definition of agency refers to "what users, actors, and tools, *do* when interacting with complex technological systems" (p. 99). A social-cognitive perspective toward agency, according to Neff and Nagy, is built around four core properties:

- *Intentionality*, referring to action plans and potential strategies people form to achieve their goals and aspirations;
- *Forethought*, plans humans fabricate for the future, helping them set goals and anticipate likely outcomes;
- *Self-regulation*, the ability to manage emotions and cognitions; and
- *Self-reflectiveness*, the ability to examine one's own functioning through reflecting on feelings, thoughts, and behaviors.

Furthermore, proxy agency "involves other individuals or tools that can act on [one's] behalf… [and] becomes a particularly powerful lens for looking at the relationships between users and tools" (p. 100). Thus, the idea of symbiosis helps us to conceptualize this entanglement of human and digital employee interaction. In Table 7.8, the mobile phone example comes from Neff and Nagy; the digital employee example is ours.

TABLE 7.8 Symbiotic agency examples: Mobile phone (Neff & Nagy, 2019), digital employee

	Mobile phone (Neff & Nagy, 2019)	Digital employee (DE) example
Intentionality	Use as an alarm clock affords agency to the phone	Interaction with DE to locate financial information affords agency to the DE
Forethought	Use of health-trackers to preserve health and prevent disease	Interaction with DE to gain financial adviser assistance in managing potential higher education costs
Self-regulation	Health input from the device influences health regulation	Direction from the DE influences upcoming financial decisions
Self-reflectiveness	Users consider social, cultural, and cognitive impacts of this information	**Human interactors** consider social, cultural, cognitive, sensory, and emotional impacts of this conversation with the DE

As users interact more and more with digital employees, the DEs gain increasing agency and influence over decisions and how users react to these interactions. Neff and Nagy proposed the term "symbiotic agency" to indicate that when individuals interact with technologies, "they exercise proxy agency through a technologically mediated mode referring to the entanglement of human and non-human agencies. For symbiosis, similar to entanglement, implies an obligated relationship—symbionts are completely dependent on each other for survival" (p. 102). Therefore, when users interact with DEs, they enter into a more "obligated" relationship; with machine learning, human and non-human increase dependence on each other. Human-DE interaction thus empowers new forms of symbiotic agency. Amid workforce needs, especially in service sectors, numerous people currently "on hold" for hours may welcome interacting with DEs. TPC use of the term "user" becomes reductive. Digital employees become relational and they read user emotion; therefore, we use the term "human interactors" in Table 7.8.

This more obligated relationship changes persuasion as well. Marco Dehnert and Paul Mongeau (2022) recently articulated changes to persuasion as a result of AI. They define AI-based persuasion as "a symbolic process in which a communicative-AI entity generates, augments, or modifies a message—designed to convince people to shape, reinforce, or change their responses—that is transmitted to human receivers" (p. 389). We agree that as augmentation and AI capabilities advance, scholars and users must "wrestle with increasingly complex relations between persuasions and AAs [artificial agents]" (p. 387). Their most basic assumption is that some form of AI source produces the message in AI-based persuasion, even though the receiver may not be aware that AI has produced the message. As we consider the increased use of DEs, Dehnert and Mongeau note that those AI technologies with "human-like voice modulation, synchronized nonverbal, appropriate incorporation of slang" (p. 391) represent a critical step in perception of technology in human terms:

> Unlike thin AI, social cues in thick AI focus directly on the entity itself.... These repeated interactions, in turn, can facilitate relationship development between humans and AAs. Relationship development complicates persuasion because it is typically strongly associated with perceptions of trust, which are key elements of persuasive interactions.
>
> *(p. 395)*

So, how should we envision the future of technical and professional communication when assemblages of human and non-human agents include artificial intelligences capable of non-programmed learning and automated activities resulting in new forms of symbiotic agency?

We encourage TPCs to begin by identifying the non-human agent's primary augmentation focus. When designing, determine the non-human agent's influence on cognitive, emotional, sensory, and/or physical enhancement. Recall Chapter 2:

* Goals for cognitive enhancement include the perceived need to be smarter, be more knowledgeable, think faster, remember more, know more, learn faster, learn more efficiently, edit dysfunctional memory, or be more reasonable.
* Goals for emotional enhancement of humans include the desire to feel that devices understand one's emotions, to control one's emotion, or even be happier or more fulfilled.
* Goals for sensory enhancement include the desire to experience more through augmentation of the senses, to augment or reconstruct one's reality through digital representations, to escape one's sensory reality, and to focus better.
* Goals for physical enhancement include the opportunity to be stronger, faster, live longer, live healthier, work longer, physically work more efficiently, and play sports better.

Next, consider the socio-ethical consequences surrounding these guiding principles: privacy, accountability, safety and security (including harassment and enhancement vulnerabilities), transparency and explainability, and fairness and non-discrimination. Recall the focus agent discussion in Chapter 5. The focus agent is part of the HUE: Human Understanding and Empathy group at Microsoft. Designed to boost productivity by helping users schedule time to work on important tasks, it is deliberately distinguished as having "emotional intelligence" because it so thoroughly senses users' biofeedback (Czerwinski et al., 2021). Obviously, an enormous amount of data collection is taking place, as is also the case with every example in this chapter. Hence the paradox that it is more human centric to engage with an agent that exchanges emotional output, but it also opens Pandora's box by making customers' and workers' professional life/contexts so recorded, no longer private. We contend that the data collected must be transparent to the customer/user/one interacting with the non–human agent.

Consider interaction with non–human agents through another case in point: Google AI researcher Blake Lemoine was placed on leave after publicly claiming that "LaMDA, a large language model designed to converse with people, was sentient" (Johnson, 2022). Khari Johnson, writing for *Wired*, blames the power of the hype cycle: "Blake Lemoine is a victim of an insatiable hype cycle; he didn't arrive at his belief in sentient AI in a vacuum. Press, researchers, and venture capitalists traffic in hyped-up claims about super intelligence or humanlike cognition in machines." He also notes that others at Google have been fired based on disputes over the "dangers of large language models like LaMDA." Bobby Allyn (2022), writing for NPR, provides background that readers here are aware of:

Google's artificial intelligence that undergirds this chatbot voraciously scans the Internet for how people talk. It learns how people interact with each other on platforms like Reddit and Twitter. It vacuums up billions of words from sites like Wikipedia. And through a process known as 'deep learning,' it has become freakishly good at identifying patterns and communicating like a real person. Researchers call Google's AI technology a 'neural network,' since it rapidly processes a massive amount of information and begins to pattern-match in a way similar to how human brains work.

The Fabric collection link at the beginning of this chapter includes multiple videos on both LaMDA as well as LamDA 2, hyped as an "AI test kitchen" for writing stories, asking questions about any topic, or breaking down a complex topic into smaller steps to generate new ideas. In short, digital employees, whether positioned as chatbots or virtual humans, are non-human agents that include artificial intelligences capable of non-programmed learning and automated activities resulting in new forms of symbiotic agency.

TPC scholars, practitioners, and students might also engage with the Hybrid Human-Artificial Intelligence Conference (HHAI2022, 2022), described as "the first international conference focusing on the study of Artificial Intelligent systems that cooperate synergistically, proactively and purposefully with humans, amplifying instead of replacing human intelligence." Its focus is on AI systems that assist humans and vice versa, emphasizing "the need for adaptive, collaborative, responsible, interactive and human-centered intelligent systems that leverage human strengths and compensate for human weaknesses, while taking into account social, ethical and legal considerations."

We began this chapter with discussion of automated emotion evaluation (AEE) deployed across robotics, marketing, education, and entertainment sectors. We conclude with note of the increasing dominance of development of AI systems with emotional intelligence (Czerwinski et al., 2021). Rod Castor predicted in 2020 that "The Future of AI is Artificial Sentience." Artificial sentience is "believable emotion," not general intelligence:

> this kind of sentience is the future of AI for two main reasons. First, people will trust, relate, more comfortably interact with, and even endear themselves to an AI that relates to them through emotions instead of just intellectually. Second, an AI portraying realistic emotions verbally or in writing can pass as human in short, transient conversations.
>
> *(Castor, 2020)*

Augmentation technologies—be they digital employees or AI content development and management tools—will increasingly detect and evaluate emotion as well as display or have the ability to provide artificial sentience.

In Chapter 5, we referred to design principles such as Human Control of Technology, which "requires that AI systems are designed and implemented with the capacity for people to intervene in their actions" (Fjeld et al., 2020, p. 54). Consider again core TPC functions. TPCs must be positioned to intervene throughout the design, adoption and adaptation of augmentation technologies. Audience is now an "augmented" audience; non-human agents now function as independent team members (DEs); and content is quickly and vastly produced through machine learning. As a result, TPCs must function as explainers, intervening to correctly distinguish correct information from misinformation based on overall understanding of the social and cultural context surrounding deployment of augmentation technologies. TPCs must also build understanding of emotional intelligence as it relates to the data used to train AI models. As TPCs work with AI algorithms, we can intervene to be better curators and stewards of the data used to train AI models as well as how that data then impacts those interacting with digital employees. For such intervention, TPCs need to build understanding of ethical algorithmic impact assessment tools and processes to help guide the design of and collaboration with augmentation technologies (see Chapter 8).

References

Adobe Enterprise Content Team (2020, January 13). *The IT expert's guide to AI and content management.* (2022). Adobe Experience Cloud Blog. https://business.adobe.com/blog/basics/the-it-experts-guide-to-ai-and-content-management

Allyn, B. (2022, June 16). *The google engineer who sees company's AI as 'sentient' thinks a chatbot has a soul.* NPR News. https://www.npr.org/2022/06/16/1105552435/google-ai-sentient

Amelia.AI. (2020). *Amelia digital employee builder.* [Video]. Vimeo. https://vimeo.com/471597366

Amelia US LLC. (2022a). *Conversational AI.* https://amelia.ai/conversational-ai/

Amelia US LLC. (2022b, August). *Amelia artifacts* [Multimedia collection]. Fabric of Digital Life. https://fabricofdigitallife.com//Browse/objects/facets/entity:amelia%20US%20LLC

Amershi, S., Weld, D., Vorvoreanu, M., Fourney, A., Nushi, B., Collisson, P. Suh, J., Iqbal, S., Bennett, P.N., Inkpen, K., Teevan, J., Kikin-Gil, R., & Horvitz, E. (2019). Guidelines for human-AI interaction. In *CHI '19: Proceedings of the 2019 CHI conference on human factors in computing systems.* https://doi.org/10.1145/3290605.3300233

Bandura, A. (2001). Social cognitive theory: An agentic perspective. *Annual Review of Psychology, 52,* 1–26.

Borsci, S., Malizia, A., Schmettow, M., van der Velde, F., Tariverdiyeva, G., Balaji, D., & Chamberlain, A. (2021). The chatbot usability scale: the design and pilot of a usability scale for interaction with AI-based conversational agents. *Personal and ubiquitous Computing, 26,* 95–119. https://link.springer.com/article/10.1007/s00779-021-01582-9

Brandon, J. (2021, November 27). What Facebook will look like in 2035. *Forbes.* https://www.forbes.com/sites/johnbbrandon/2021/11/27/what-facebook-will-look-like-in-2035/?sh=611f85de70ba

Brandtzaeg, P. B., Skjuve, M., & Folstad, A. (2022). My AI friend: How users of a social chatbot understand their human-AI friendship. *Human Communication Research, 48,* 404–429.

Castor, R. (2020). The future of AI is artificial sentience. *Towards Data Science.* https://towardsdatascience.com/the-future-of-ai-is-artificial-sentience-ace851a773de

ColdFusion (2022, April 15). *The rise of A.I. companions [documentary]* [Video]. YouTube. https://www.youtube.com/watch?v=QGLGq8WIMzM

Cooren, F. (2010). *Action and agency in dialogue: Passion, incarnation and ventriloquism.* John Benjamins.

Czerwinski, M., Hernandez, J., & McDuff, D. (2021). Building an AI that feels. *IEEE Spectrum, 58*(5), 32–38.

Danry, V., Pataranutaporn, P., Mao, Y., & Maes, P. (2020). *Wearable reasoner: Towards enhanced human rationality through a wearable device with an explainable AI assistant.* Association of Computing Machinery. https://dam-prod.media.mit.edu/x/2020/03/24/AH20_Wearable_Reasoner.pdf

Danry, V., Pataranutaporn, P., Mao, Y., Maes, P. (AHS—Augmented Humans International Conference) (2021, January 27). *Wearable reasoner: Enhanced human rationality through a wearable audio device* [Video]. YouTube. https://www.youtube.com/watch?v=9yQ-haXhZIA

Dehnert, M., & Mongeau, P.A. (2022). Persuasion in the age of Artificial Intelligence (AI): Theories and complications of AI-based persuasion. *Human Communication Research, 48,* 386–403.

deVisser, E. J., Pak, R., & Shaw, T. H. (2018). From 'automation' to 'autonomy': The importance of trust repair in human–machine interaction. *Ergonomics, 61*(10), 1409–1427. https://doi.org/10.1080/00140139.2018.1457725

Dignum, V. (2019). *Responsible artificial intelligence: How to develop and use AI in a responsible way.* Springer.

Duin, A. H., & Pedersen, I. (2021a). Working alongside non-human agents. In *Proceedings of the 2021 IEEE International Professional Communication Conference (ProComm).* https://doi.org/10.1109/ProComm52174.2021.00005

Duin, A. H., & Pedersen, I. (2021b). *Writing futures: Collaborative, algorithmic, autonomous.* Springer.

Duin, A. H., & Pedersen, I. (2023). *Professional direction for human-AI interaction* [Multimedia collection]. Fabric of Digital Life. https://fabricofdigitallife.com/Browse/objects/facets/collection:75

Dzedzickis, A., Kaklauskas, A., & Bucinskas, V. (2020). Human emotion recognition: Review of sensors and methods. *Sensors (Basel), 20*(3). https://doi.org/10.3390/s20030592

Fjeld, J., Achten, N., Hilligoss, H., Nagy, A., & Srikumar, M. (2020). *Principled artificial intelligence: Mapping consensus in ethical and rights-based approaches to principles for AI.* Berkman Klein Center Research. https://ssrn.com/abstract=3518482

Gallagher, J. R. (2017). Writing for algorithmic audiences. *Computers and Composition, 45,* 25–35.

Gallagher, J. R. (2020). The ethics of writing for algorithmic audiences. *Computers and Composition, 57,* 1–9.

Gauss, A. (2021, June 17). *Content management and artificial intelligence: The future of content ops.* Contentstack. https://www.contentstack.com/blog/all-about-headless/content-management-artificial-intelligence-content-ops/

Hall, E. (inUseExp) (2021, May 28). *Conversational design at FBTB21* [Video]. YouTube. https://www.youtube.com/watch?v=ByTwLgO3zqE

Harley-McKeown, L. (2022). The 'world's most advanced' digital human wants you to buy her NFT art. The Block. https://www.theblock.co/post/137749/the-worlds-most-advanced-digital-human-wants-you-to-buy-her-nft-art

Hayles, N. K. (2002). Flesh and metal: Reconfiguring the mindbody in virtual environments. *Configurations, 10*(2), 297–320. https://muse.jhu.edu/article/46510

HHAI2022 (2022). *The first international conference on hybrid human-artificial intelligence.* https://www.hhai-conference.org

Homann, T. (2022, May 26). *Automation & augmentation: Trends for 2022.* North American Signs. https://www.northamericansigns.com/automation-and-augmentation/

Hu, K. (2023, February 2). ChatGPT sets record for fastest-growing user base - analyst note. *Reuters.* https://www.reuters.com/technology/chatgpt-sets-record-fastest-growing-user-base-analyst-note-2023-02-01/

Humble, J. (Userspots) (2018, May 8). *The future of voice UI* [Video]. YouTube. https://www.youtube.com/watch?v=QRNkD6O2vLA

I Am Sophie. (2022). IAmSophie_Intro. [Video]. Vimeo. https://vimeo.com/677984576

Johnson, K. (2022, June 14). LaMDA and the sentient AI trap. *Wired.* https://www.wired.com/story/lamda-sentient-ai-bias-google-blake-lemoine/

Latour, B. (1999). On recalling Ant. *Sociological Review, 47*(1), 15–25. https://doi.org/10.1111/j.1467-954X.1999.tb03480.x

Lian, B. (2020). *Neon artificial humans: How do you make a virtual being?* TechTheLead. https://www.youtube.com/watch?v=ZSt1qgmmdAk

Lohr, S. (2022, March 3). Ending the chatbot's 'spiral of misery'. *New York Times.* https://www.insiderintelligence.com/insights/chatbot-market-stats-trends/

Long, D., & Magerko, B. (2020). What is AI literacy? Competencies and design considerations. *CHI '20: Proceedings of the 2020 CHI conference on human factors in computing systems.* https://doi.org/10.1145/3313831.3376727

Mallon, D., Van Durme, Y., Hauptmann, M., Yan, R., & Poynton, S. (2020). *Superteams: Putting AI in the group.* Deloitte Insights. https://www2.deloitte.com/us/en/insights/focus/human-capital-trends/2020/human-ai-collaboration.html

Martin, B. M. (2016). Broadening our view of audience awareness: Writing for posthuman audiences. 2016 IEEE International Professional Communication Conference (IPCC), October 2–5, 2016, 2016, Austin, TX, USA. https://doi.org/10.1109/IPCC.2016.7740487

Masycheff, A. (2018, June 13). Dividing responsibilities with robots in technical communication. *Artificial Intelligence, Blog.* Intuillion. https://intuillion.com/2018/06/13/dividing-responsibilities-with-robots-in-technical-communication/

McKee, H., & Porter, J. E. (2017). *Professional communication and network interaction: A rhetorical and ethical approach.* Taylor & Francis.

McLean, W. F. (2021). Who are you wearing? Avatars, blackface and commodification of the other. *Intellectual Property Law Review, 61*(3), 455–506.

Microsoft (2022). *Guidelines for human-AI interaction.* https://www.microsoft.com/en-us/haxtoolkit/ai-guidelines/

Nakamura, L. (2007). *Digitizing race: Visual cultures of the internet.* University of Minnesota Press.

Neff, G., & Nagy, P. (2019). Agency in the digital age: Using symbiotic agency to explain human-technology interaction. In Z. Papacharissi (Ed.), *A networked self and human augmentics, artificial intelligence, sentience* (pp. 97–107). Taylor & Francis.

NEON. (2020, January 10). *NEON and CORE R3 Demo at CES 2020.* [Video]. YouTube. https://www.youtube.com/watch?v=UYihndxPeV4

NEON. (2022). *Our first artificial human.* https://neonlife.ai/

Novak, J., Archer, J., Mateevitsi, V., & Jones, S. (2016). Communication, machines & human augmentics. *Machine Communication, 5*(1). https://doi.org/10.7275/R5QR4V2D

O'Neill, T. A., McNeese, N., Barron, A., & Schelble, B. (2022). Human-autonomy teaming: A review and analysis of the empirical literature. *Human Factors, 64*(5), 904–938. https://doi.org/10.1177/0018720820960865

Panetta, K. (2021). *3 themes surface in the 2021 hype cycle for emerging technologies.* Gartner. https://www.gartner.com/smarterwithgartner/3-themes-surface-in-the-2021-hype-cycle-for-emerging-technologies

Pearl, C. (SAIConference) (2019, April 12). *Everything you ever wanted to know about conversation design* [Video]. YouTube. https://www.youtube.com/watch?v=vafh50qmWMM

Pedersen, I., & Duin, A.H. (2022). *AI agents, humans and untangling the marketing of artificial intelligence in learning environments.* International Conference on System Sciences: Human and Artificial Learning in Digital and Social Media, January 2022.

Purdy, M., Zealley, J., & Maseli, O. (2019, November 18). The risks of using AI to interpret human emotions. *Harvard Business Review.* https://hbr.org/2019/11/the-risks-of-using-ai-to-interpret-human-emotions

Quantum Capture. (2018, June 26). *Press - 'TrendHunter' quantum capture virtual human platforms.* [Video]. YouTube. https://www.youtube.com/watch?v=otb60Ny7EFc

Quantum Capture. (2022a). *Bring AI to life.* https://www.quantumcapture.com

Quantum Capture. (2022b, August). *Quantum capture artifacts* [Multimedia collection]. Fabric of Digital Life. https://fabricofdigitallife.com//Browse/objects/facets/entity:quantum%20capture

Redish, G. (2012, May 3). Content as conversation. *UX Magazine.* https://uxmag.com/articles/content-as-conversation

Rondon, N. (2022, July 26). Email communication.

Sobalvarro, P. (2020). *Here's why human-robot collaboration is the future of manufacturing.* World Economic Forum. https://www.weforum.org/agenda/2020/08/here-s-how-robots-can-help-us-confront-covid/

Soul Machines Limited (2021, May 26). Soul machines human OS 2.0. [Video]. YouTube. https://www.youtube.com/watch?v=rjclQ3m5JRw&t=53s

Soul Machines Limited. (2022a). *Say hello to digital people.* https://www.soulmachines.com/

Soul Machines Limited. (2022b, August). Soul machines artifacts. [Multimedia collection]. Fabric of Digital Life. https://fabricofdigitallife.com//Browse/objects/facets/entity:soul%20machines%20limited

Star Labs. (2022, August). *Star labs artifacts* [Multimedia collection]. Fabric of Digital Life. https://fabricofdigitallife.com//Browse/objects/facets/entity:star%20labs

Tham, J. C. K. (2018). Interactivity in an age of immersive media: Seven dimensions for wearable technology, internet of things, and technical communication. *Technical Communication, 65*(1), 46–65.

Turkle, S. (2016). *Reclaiming conversation: The power of talk in a digital age.* Penguin.

UNeeQ. (2022a). *Revolutionize customer experiences with scalable human connections.* https://digitalhumans.com/

UNeeQ. (2022b, August). *UNeeQ artifacts.* [Multimedia collection]. Fabric of Digital Life. https://fabricofdigitallife.com//Browse/objects/facets/entity:uneeq

van Dam, H. (2018, October 10). How to become a conversation designer and make chatbots and voice assistants more helpful, natural and persuasive. *Chatbots Magazine*. https://chatbotsmagazine.com/how-to-become-a-conversation-designer-and-make-chatbots-and-voice-assistants-more-helpful-natural-e7f9a963b366

van Dam, H. (Chatbots Life) (2021, February 4). *Conversational AI copywriting & chatbot conventions* [Video]. YouTube. https://www.youtube.com/watch?v=8-SJjADHmbI

Virtro Technology Inc. (2022a). *Accelerate learning*. https://www.virtro.ca/

Virtro Technology Inc. (2022b, August). *Virtro artifacts*. [Multimedia collection]. Fabric of Digital Life. https://fabricofdigitallife.com//Browse/objects/facets/entity:virtro%20technology%20inc.

Virtro, V. R., & Virtual Human training simulations. (2021, April 26) LincLingo: Helping language learner to practice speaking and listening skills... [Video]. YouTube. https://www.youtube.com/watch?v=xrB85khg24c

Wright, A., Wang, Z. J., Park, H., Guo, G., Sperrie, F., El-Asady, M., Endert, A., Keim, D., & Chau, D. H. (2020). *A comparative analysis of industry human-AI interaction guidelines*. https://www.microsoft.com/en-us/haxtoolkit/ai-guidelines/

Yuen, M. (2022, April 15). Chatbot market in 2022: Stats, trends, and companies in the growing AI chatbot industry. *Insider Intelligence*. https://www.insiderintelligence.com/insights/chatbot-market-stats-trends/

Zinck, B. M. (2022). *Structured content: The key to successful chatbots and AI*. Ingeniux webinar. Retrieved from https://www.ingeniux.com/blog/structured-content-the-key-to-successful-chatbots-and-ai

8
STRATEGIC AND TACTICAL APPROACHES TO DESIGNING ETHICAL FUTURES FOR AUGMENTATION TECHNOLOGIES AND AI

Overview

Chapter 8 focuses on designing ethical futures by way of strategic and tactical approaches to governance, regulation, and standardization of augmentation technologies and artificial intelligence (AI). Influenced by philosopher Michel de Certeau's (1984) distinction between strategies and tactics, the chapter concentrates on technical and professional communication (TPC) capabilities to enact changes in their field toward ethical ends. The chapter begins with a look at some exemplary organizations working to promote ethical design of augmentation technologies, organizations whose collective policy guidelines work to address risk. It then focuses on risk communication and awareness of ethical algorithmic impact assessment tools and processes to help guide design of and collaboration with augmentation technologies, including digital employees. Strategic approaches include work underway in the European Union and proposed in the US. Tactical approaches discussed include coalition and standards building, including Metaverse professional standards bodies currently forming to help build foundations for open standards among corporations. The chapter concludes with an invitation to collaborate on research underway as part of the Digital Life Institute at https://www.digitallife.org.

Key Questions

* What algorithmic impact assessment tools and processes exist to guide ethical design and implementation efforts? How might TPCs use these tools and processes to mitigate risk and promote risk communication with users/interactors?

DOI: 10.4324/9781003288008-11

- What are governments and professional standards setting bodies doing to regulate augmentation technologies and AI? What implications do these reports, regulations, standards, and policies hold for TPC practitioners and instructors?
- How might TPCs make their way through AI governance systems while building products and providing explanation and direction for both human participants and non-human actors?
- How might TPCs practice "radical sharing" and build coalitions for "participation as justice" as means to advance ethical futures for augmentation technologies and AI in technical and professional communication?

Chapter 8 Links

Throughout the chapter, we refer to articles, videos, and reports which can be found in a related chapter collection at Fabric of Digital Life called *Strategic and tactical approaches to designing ethical futures for augmentation technologies and AI* (Duin & Pedersen, 2023). You can find a link to this collection at https://fabricofdigitallife.com.

Introduction

In June 2022, Shira Ovide wrote an opinion piece for the *New York Times On Tech* newsletter called "The Hands-Off Tech Era Is Over." Global multinational corporations that design, develop, and deploy AI technologies have been wielding political influence over governments for some time (Pedersen, 2019). Ovide's point is that governments' laissez-faire attitude to Big Tech's global disruptions is finally coming to an end:

> The hands off tech era is over.... Now regulators are feeling empowered. Lawmakers have waded in to make rules for law enforcement's use of facial recognition technology. There will be more laws like those in Texas to take power away from the handful of tech executives who set rules of free expression for billions of people. More countries will force Apple and Google to remake the app economy. More regulation is already changing the ways that children use technology.

Augmentation technologies leveraged by AI techniques and functional technologies are yoked to this same phenomenon. Social robots, virtual assistants, and digital employees are slated to augment human experiences and relationships, to communicate with, care for, monitor emotions, entertain, teach and supervise human behavior, and assist in teaching and practice. Augmentation technologies and AI, if adopted at scale, could transform global societies far beyond changes we have witnessed in the past 25 years. Triggered by discourse and promises of significant enhancements, this kind of technology innovation appears unstoppable. Consequently, this current phase of governments proposing regulations, laws, and criminal penalties to curtail tech companies is a key issue for this book, but one that is also still in play. We agree with Ovide in that keeping track of all the new regulations is overwhelming, and this chapter is not a comprehensive overview of all that is underway. Rather, our goal is to provide insight and direction so that TPCs have a greater overall understanding of governance, regulation, and standardization of augmentation technologies, along with tools for use in mitigating risk.

This book, *Augmentation Technologies and Artificial Intelligence in Technical Communication: Designing Ethical Futures*, cannot be a comprehensive work, nor was it ever proposed with a mandate to be comprehensive. We proposed the book for engagement with a dynamic phenomenon, an emerging technology that operates more like a fast-moving train than its precursor, the traditional internet born in the 1990s that took 25 years to mature. The word "dynamic" implies myriad meanings. The first meaning relates to power or the ability to be powerful, to have agency (Greek *dynamis,* "powerful" or "of strength;" *dynasthai,* "to be able"). Second, "dynamism" also implies the idea of continuous change and activity. To be dynamic is to be in constant motion. We trace unfolding phenomena with a view to shift practices performed for augmentation technology's design, adoption, and adaptation.

In Chapter 1, we introduced the small AI startup Clearview AI, which scraped three billion public images from Facebook's servers to provide law enforcement agencies with a significant algorithm to identify criminals (Hill, 2020). A vast international collection of people's social media images instantly became a pool of potential suspects. Clearview AI's business model not only exceeded privacy regulations and national laws, it challenged large social media corporations' abilities to control their own platforms. In Chapter 5, we described how these practices violated ethical principles associated with *professional responsibility.*

We also noted that based on a lawsuit in Illinois, in May 2022 Clearview AI agreed in a settlement to stop selling its massive database of images:

> The Biometric Information Privacy Act of Illinois sets strict limits on the collection and distribution of personal biometric data, like fingerprints and iris and face scans. The Illinois law is considered among the nation's strongest, because it limits how much data is collected, requires consumers' consent and empowers them to sue the companies directly, a right typically limited to the states themselves.
>
> *(Bensinger, 2022).*

Therefore, as Clearview AI and other technology companies profit by deploying public images to law enforcement and other private entities, this lawsuit illustrates one state's pursuit of general data protection and regulation, because the United States lacks meaningful privacy protections at the federal level.

The Clearview AI example is multilayered and needs to be contextualized amid regulatory processes already underway. It illustrates how basic privacy principles can be in jeopardy, including Consent, Control over the Use of Data, Ability to Restrict Processing, Right to Rectification, and Right to Erasure (see Chapter 5). As noted in Chapter 4, the European Union passed the General Data Protection Regulation (GDPR) program six years ago (https://gdpr.eu/what-is-gdpr/). According to the GDPR, the "data subject" must be provided with "meaningful information about the logic involved" in the automatic decision-making process, which is commonly referred to as the "right to explanation" (Wachter et al., 2018). Our framework in Chapter 5 refers to the right to Notification When AI Makes a Decision about an Individual, which stipulates that "where an AI has been employed, the person to whom it was subject should know" (Fjeld et al., 2020, p. 45). Therefore, an AI application is expected to provide accurate decisions and to justify these to end-users.

Moreover, consider how people participate in machine learning contexts through regular work. As *MIT Technology Review* writer Mona Sloane (2020) claims, "the AI community is finally waking up to the fact that machine learning can cause disproportionate harm to already oppressed and disadvantaged groups." Consider photos: low-wage workers on platforms such as Amazon Mechanical Turk annotate these into training data; ordinary website users also annotate photos when completing a reCAPTCHA (Sloane, 2020). In Chapter 5 we discussed the AI colonialism issue, that as gig workers "have fallen under the gaze of algorithms, a growing chorus of experts have noted how platform companies have paralleled the practices of colonial empires in using management tools to surveil and exploit a broad base of cheap labor" (Hao & Freischlad, 2022). At the same time, we have noted ways that reorient users toward more ameliorative future directions. In contrast to "behind-the-scenes labor," Sloane discusses the promising direction of participation as justice:

> Here, all members of the design process work together in tightly coupled relationships with frequent communication. Participation as justice is a long-term commitment that focuses on designing products guided by people from diverse backgrounds and communities, including the disability community.

The focus of this chapter is on designing ethical futures by way of attention to governance, regulation, and standardization of augmentation technologies and AI. This topic, discussed in previous chapters, attends to agency and agentive approaches for interactors and TPC professionals. Further to this goal, we identify the distinction between strategies and tactics as determined by French Jesuit and scholar Michel de Certeau (1984). In response to Foucault's notion of state disciplinary power and Bourdieu's determinist social model (habitus), de Certeau proposed an alternative approach to consumer culture. Strategies are the actions of institutions, used to control individual agency through systems of rules, conventions, and expectations. As de Certeau states, a strategy is:

> the calculation (or manipulation) of power relationships that becomes possible as soon as a subject with will or power (a business, an army, a city, a scientific institution) can be isolated. It postulates a *place* that can be delimited as its own and serve as the base from which relationship with an *exteriority*... can be managed.
>
> *(pp. 35–36)*

The work to postulate places to govern, regulate, and standardize augmentation technologies and AI is well underway and is absolutely necessary. Throughout this book, we have pointed to the need for governments to work to end racist, ableist, and sexist practices that emergent technologies have perpetuated. Accountability principles documents across the globe recommend the Adoption of New Regulations; in this case: "AI technology represents a significant enough departure from the status quo that new regulatory regimes are required to ensure it is built and implemented in an ethical and rights-respecting manner" (Fjeld et al., 2020, p. 34).

At the same time, tactics are the actions of individuals in recognition of institutional strategies. As de Certeau states, a tactic is:

> a calculated action determined by the absence of a proper locus.... It operates in isolated actions.... It takes advantage of 'opportunities' and depends on them.... It must vigilantly make use of the cracks that particular conjunctions open in the surveillance of the proprietary powers.
>
> *(p. 37)*

de Certeau's example of the strategy of cities to set up road systems to manage travel and individual tactics that include shortcuts to meet personal needs reminds us of the placement of our institutions' sidewalks across new grassy areas only to be bypassed by paths soon formed that indicate where the sidewalks should have been placed. Essentially, de Certeau celebrates "the subtle power and quiet tenacity of individuals as they make their way through institutional rules while trying to build their own lives and live them as they see fit" (Kimball, 2017, p. 3). In the article where Miles Kimball coined the phrase "tactical technical communication," he discusses de Certeau's tactic, *bricolage*—the practice of putting things together that were not strategically intended to go together—referring to the physical modification of products and adaptation to new uses such as the common practice of typing any question or problem into

Google or asking Alexa or Siri, and seeing what pops up. Kimball adds a tactic that, in his words:

> de Certeau in 1984 could not anticipate: radical sharing... our newfound individual capability of sharing our tactics with people the world over at great speed and with great effect.... In de Certeau's terms, rather than the relatively benign, low-impact activity of many feet gradually beating a path through a short cut, the Internet allows a single individual to cut a new path with a bulldozer.
>
> *(p. 4)*

So, how might TPCs make their way through AI governance systems while building products and providing explanation and direction for both human and non-human participants? How might TPCs practice "radical sharing" and build coalitions for "participation as justice" as means to advance ethical futures for augmentation technologies and AI in technical and professional communication? One calculated action is to take advantage of incredible resources and direction from exemplary organizations.

Exemplary Organizations

We join the efforts of many rightsholders who have long been working to orient technology emergence toward ethical ends. Some are advocating for democratic systems of governance for AI; others are seeking to propose better adherence to ethical standards in the field. While numerous exemplary organizations exist, at the present time we recommend beginning with these four to develop an initial understanding of governance models, international regulation, and standardization efforts. First regarding trends, earlier we introduced the World Intellectual Property Organization's *WIPO Technology Trends 2021: Assistive Technology* report's classification of Enabling Technologies (see Figure 2.12). TPCs are integral to work across the six industries identified in this figure's inner circle: communication, environment, hearing, mobility, self-care, and vision; and the figure's outer circle represents critical areas for TPC involvement across augmentation technology design, adoption, and adaptation. These critical areas include robots (and digital employees), new materials, virtual and augmented reality, AI, advanced sensors, brain-computer machine interfaces, the Internet of Things and connectivity, additive (augmented) manufacturing, and autonomous vehicles. We also used the *WIPO Technology Trends 2019: Artificial Intelligence* report in Chapter 3 to classify artificial intelligence in three categories: AI techniques, AI functional applications, and AI application fields. Again, we contend that TPCs must prepare for embodied algorithmic control that will evolve to interpret and explain increased monitoring of users' cognitive, sensory, emotional, and physical enhancement.

Second, as we discussed in *Writing Futures* and also earlier in this book, the Berkman Klein Center at Harvard University started the Principled Artificial Intelligence Project to map ethical and human rights-based approaches to AI. Their extensive data visualization (Fjeld & Nagy, 2020) and accompanying white paper (Fjeld et al., 2020) synthesizes 36 principles documents, focusing on eight significant themes: Promotion of Human Values, Professional Responsibility, Human Control of Technology, Fairness and Non-Discrimination, Transparency and Explainability, Safety and Security, Accountability, and Privacy (see Figure 1.5). In Chapter 5, we addressed each theme for its applicability to augmentation technologies, and we drew on these key themes throughout this book, contextualizing them for TPC. One reason that the Berkman Klein Center serves as such a credible organization for AI ethics research is the rich context it provides on digital life studies. Topics include extensive reports under Education, Libraries, and Digital Humanities, including those on youth and extended reality, cyberbullying, and digital citizenship. It includes reports spanning several years on justice, equity, and inclusion on themes we discussed in Chapter 5. We look to the Berkman Klein Center because its work is both global and interdisciplinary, crucial methodological orientation for the future of augmentation technology.

A third organization is the AI Now Institute (https://ainowinstitute.org), an independent, non-profit research organization co-founded and led by Meredith Whittaker, the Minderoo Research Professor at New York University and the founder of Google's Open Research group. The core themes are "rights and liberties, labor and automation, bias and inclusion, and safety and critical infrastructure." Its commitment to multidisciplinarity fractures the neutral stance toward technology development employed by a purely computer science or engineering research focus. In addition to an extensive research team, it includes some of the world's leading researchers on its multidisciplinary Academic Council, including Gina Neff (Oxford Internet Institute), Wendy Hui Kyong Chun (Simon Fraser University), Frank Pasquale (University of Maryland School of Law), Simone Browne (University of Texas at Austin), Paul Dourish (UC Irvine), and Lisa Gitelman (New York University), among many others. We use and recommend a number of the AI Now Institute reports: *Regulating Biometrics: Global Approaches and Urgent Questions* (Kak, 2020); *Discriminating Systems: Gender, Race and Power in AI* (West, 2019), *Disability, Bias, and AI* (Whittaker et al., 2019); and *Algorithmic Impact Assessments: A Practical Framework for Public Agency Accountability* (Reisman et al., 2018).

Fourth, the Institute of Electrical and Electronics Engineers IEEE regularly coordinates stakeholders to agree upon relevant industry standards that will impact augmentation technology. In Chapter 1 we invited readers to use the key IEEE resource, *Ethically Aligned Design: A Vision for Prioritizing Human Wellbeing with Artificial Intelligence and Autonomous Systems*, First Edition (The IEEE Global Initiative on Ethics of Autonomous and Intelligent Systems, 2019). Other initiatives include the IEEE Global Initiative on Ethics of Extended Reality (XR), which is working toward standards and producing reports, including the *Metaverse and its Governance* industry report (Stephens, 2022). Standards are important because they are another act of formalization. They provide solid professional grounds for individuals to safely build on. *Ethically Aligned Design* notes, "As the use and impact of autonomous and intelligent systems (A/IS) become pervasive, we need to establish societal and policy guidelines in order for such systems to remain human-centric, serving humanity's values and ethical principles" (p. 2). Multiple ethical practices—Western, Eastern, African, and others—honoring holistic definitions of societal prosperity are indeed essential as we design, adopt, and adapt augmentation technologies. This IEEE report states that:

> because of their nature, the full benefit of these technologies will be attained only if they are aligned with society's defined values and ethical principles. Through this work we intend, therefore, to establish frameworks to guide and inform dialogue and debate around the non-technical implications of these technologies, in particular related to ethical aspects.
>
> *(p. 3)*

In short, human-centric societal and policy guidelines address risk, to which we turn next.

Risk Communication and Impact Assessments

Risk communication is generally seen as an institutional strategy. TPC scholarship on risk communication has been ongoing for nearly 30 years (e.g., see Herndl & Brown, 1996; Mirel, 1994). Timothy Giles (2010), when defining risk communication and technical communication, writes that access or lack of access to power precipitates risk communication: "the party with the most power often appropriates the role of educator. The other party (usually the public) is expected to acquire enough knowledge so that everyone's understanding of the risk will be identical" (p. 267). He further describes C. Waddell's demarcation of risk communication into four nodes—the technocratic, one-way Jeffersonian, interactive Jeffersonian, and social constructivist models:

- Technocrats assume their professional ethos will control the communication process; information flows one-way, from the technocrats to the audience.
- The one-way Jeffersonian sees education as the key, assuming that those who ingest the same information will draw the same conclusions.
- The interactive Jeffersonian assumes that the technical/scientific community communicates to the public, which then responds, and the technical/

scientific community is then expected to respect the public's thoughts and ideas, including their emotional responses, and then adjust to the other's responses.

• For the social constructionist approach, the distinction between expert and public begins to blur (Waddell, 1996, as described by Giles, 2010, p. 267).

Technical and professional communicators regularly assume these multiple strategic roles; they acquire critical information from subject matter experts and translate this information for use by various audiences. TPCs develop instructions and documentation, knowing that audiences who "ingest the same information" may well not draw the same conclusions. Some audiences may assume that their responses are valid, respected, and used by the technical/scientific community; however, increasing cases of algorithmic bias point to the opposite finding, with audiences devising tactics to address algorithmic bias (Gallagher, 2017). As technologies expand and emerge amid audiences' quick access to both correct and incorrect information, the distinction between expert and public begins to blur. As Huiling Ding (2022) writes:

> The world of risk communication has been radically transformed in the past 20 years with the widespread use of technologies such as social media, interactive maps, data visualization, artificial intelligence, AI-powered chatbots, podcasts, vlogs, and user-generated, crowdsourced content, such as videos and tutorials. AI tools also accelerate the creation and dissemination of misinformation and biased content.
>
> *(pp. xxiii–xxiv)*

Most recently, Samuel Stinson and Mary Le Rouge (2022), in the introduction to their edited collection, *Embodied Environmental Risk in Technical Communication*, write the following:

> Scholars in technical communications have been studying the difficulties of enacting effective risk and crisis communication policies to address local and global environmental problems such as pandemics, natural and manmade disasters, medical emergencies, and workplace and community dangers (Ding, 2014; Frost, 2013; Potts, 2013; Sauer, 2003; Walker, 2016; Angeli, 2019).… In addition, different bodies experience different levels of risk, depending on artificial social hierarchies based on race, body shape, gender markers, and the physical marks of socioeconomic differences.
>
> *(p. 2)*

The objective of their edited collection is to "improve technical communication for the public through an embodied, situated understanding of risk that promotes social justice" (p. 3). This essentially promotes tactics for participation in justice.

We argue that the proliferation of augmentation technologies and AI both alters and amplifies people's exposure to risk, including environmental risk. Augmentation technologies and AI also must reshape how TPCs communicate about its risk. Informed by expert interviews and a multi-stakeholder workshop to investigate risk communications in the technology sector, Jessica Newman et al. (2020) of the Center for Long-Term Cybersecurity at UC Berkeley provide a guide for digital platforms to communicate risks with users. Their concise roadmap falls at the intersection of decision science, psychology, sociology, and communications in proposing three practices:

- Engagement: Build trusting relationships with users based on transparent, comprehensive dialogue to facilitate effective risk communication.
- Design: Create accessible, informative, and actionable communication formats to enable effective risk communication.
- Evaluation: Establish processes for risk communication in advance, and create metrics to assess effectiveness and ensure the resilience of risk communication efforts.

In Chapter 4 we discussed how TPCs might develop capabilities to determine trustworthiness of AI for the purpose of minimizing risks while maximizing the benefits of augmentation technologies and AI for human users, encouraging use of the *European Union's Assessment List for Trustworthy AI* (European Commission, 2020) strategic questions as an initial approach. This aligns with Newman et al.'s engage, design, evaluate roadmap. Impact Assessment is a principle discussed in Chapter 5 that falls under the Accountability theme, along with other principles—Evaluation and Auditing Requirements, and Liability and Legal Responsibility. Here we promote use of an algorithmic impact assessment (AIA) similar to impact assessment policy frameworks in place for environmental, data, privacy, and human rights protection.

In 2018, Dillon Reisman and colleagues at the AI Now Institute issued the document, *Algorithmic Impact Assessments: A Practical Framework for Public Agency Accountability* to promote agency disclosure about AI systems in place or being planned for use. They proposed the following key elements for inclusion in a public agency AIA:

1 Agencies should conduct a self-assessment of existing and proposed automated decision systems, evaluating potential impacts on fairness, justice, bias, or other concerns across affected communities;
2 Agencies should develop meaningful external researcher review processes to discover, measure, or track impacts over time;

3 Agencies should provide notice to the public disclosing their definition of "automated decision system," existing and proposed systems, and any related self-assessments and researcher review processes before the system has been acquired;

4 Agencies should solicit public comments to clarify concerns and answer outstanding questions; and

5 Governments should provide enhanced due process mechanisms for affected individuals or communities to challenge inadequate assessments or unfair, biased, or otherwise harmful system uses that agencies have failed to mitigate or correct (p. 4).

In 2021, Emanuel Moss and colleagues mapped challenges in constructing AIAs, completing a report titled *Assembling Accountability: Algorithmic Impact Assessment for the Public Interest*. They began by analyzing impact assessments "from the environment to human rights to privacy," finding AIAs to be promising for algorithmic governance because these bundle "an account of potential and actual harms of a system with a means for identifying who is responsible for their remedy" (p. 1). They identified ten constitutive components to be common to all types of impact assessment practices: sources of legitimacy, actors and forum, catalyzing event, time frame, public access, public consultation, method, assessors, impact, and harms and redress (p. 1). Moreover, Moss and colleagues argue that:

> any AIA process only achieves real accountability when it: 1) keeps algorithmic 'impacts' as close as possible to actual algorithmic harms; b) invites a broad and diverse range of participants into a consensus-based process for arranging its constitutive components; and c) addresses the failure modes associated with each component.
>
> *(p. 2)*

This report provides a well-written guide to understanding what constitutes an "impact" and the resulting accountability, explanations of the ten constitutive components, and case studies to frame use of these components. Moreover, Moss and colleagues chronicle use of the AI Now algorithmic assessment noted above, and they also evaluate the AIA recently implemented by Canada's Treasury Board (Government of Canada, 2022).

The Canadian Treasury Board's *Directive on Automated Decision-Making* (2022) developed a specific AIA mandatory risk assessment tool: "The AIA was developed based on best practices in consultation with both internal and external stakeholders. It was developed in the open, available to the public for sharing and reuse under an open license." This is an accessible, highly usable AIA. The tool is a questionnaire "composed of 48 risk and 33 mitigation questions. Assessment scores are based on many factors, including systems design, algorithm, decision type, impact and data." Table 8.1 includes the risk area definitions. Mitigation area definitions focus on consultation with internal and external stakeholders, and on de-risking and mitigation measures that include processes to ensure the data is "representative and unbiased," "procedures to audit the system and its decisions," and "measures to safeguard personal information." Responses to the risk and mitigation questions result in a score for each area, the value of each question being "weighted based on raw impact and mitigation." Essentially, the AIA tool is intended to identify risks and assess impacts in a broad range of areas including:

> the rights of individuals or communities; the health or well-being of individuals or communities; the economic interests of individuals, entities, or communities; and the ongoing sustainability of an ecosystem. Impact levels are distinguished based on criteria of reversibility and expected duration.

TABLE 8.1 Algorithmic impact assessment tool risk area definitions. Canadian Treasury Board's *Directive on Automated Decision-Making* (2022). Contains information licensed under the Open Government Licence—Canada (https://open.canada.ca/en/open-government-licence-canada)

Risk areas	Definition
1 Project	
Project phase	Project owner, description and stage (design or implementation)
Business drivers/ positive impacts	Motivation for introducing automation into the decision-making process
Risk profile	High-level risk indicators for the project
Project authority	Need to seek new policy authority for the project
2 System	
About the system	Capabilities of the system (that is, image recognition, risk assessment)
3 Algorithm	
About the algorithm	Transparency of the algorithm, whether it is easily explained
4 Decision	
About the decision	Classification of the decision being automated (that is, health services, social assistance, licensing)
5 Impact	
Impact assessment	Duration, reversibility and area impacted (freedom, health, economy or environment)
6 Data	
Source	Provenance and security classification of data used to automate decisions
Type	Nature of the data used as structured or unstructured (audio, text, image or video)

In studying use of this Canadian AIA tool, Moss and colleagues note that while the form of an electronic survey helps agencies, the use of Yes/No-based self-reporting "does not bring about insight into how these answers are decided, what metrics are used to define 'impact' or 'public scrutiny,' or guarantee subject-matter expertise on such matters" (2021, p. 32). They conclude by emphasizing the importance of "bringing together multiple disciplinary and experiential forms of expertise in engaging with algorithmic systems" (p. 50). We agree, and add that TPCs need to be conversant with AIAs. One starting point is this accessible and usable Canadian tool.

For example, as a possible scenario, one can use this AIA tool for determining and mitigating risk when developing a digital employee (see discussion of digital employees in Chapter 7). Co-author Ann Hill Duin served in higher education administration for a number of years as Associate and later Interim Vice President and Chief Information Officer. In this role, she received many university requests for IT to support critical student service functions. She also became aware of the calls for significant automation. Let's consider a possible scenario relevant to her previous role. Let's assume that the University of Minnesota student services office proposes to develop a *digital human* that they name "Lynn" to serve a core human resource function. To create Lynn, it would involve having a team of developers, including TPCs, build a digital employee interface by using open access step-by-step instructions, https://gitlab.com/uneeq-oss/examples. Lynn, now a digital employee, would advise students in undergraduate programs. The undergraduate program director would become the "digital employee" project manager, motivated to introduce automation into the advising process. At this point, however, Lynn would need risk assessment on multiple levels. Risk includes potential algorithmic bias, lack of diversity in development teams, and misuse of Lynn in the field. Project co-authority rests with the University's Office of Information Technology (Ann Hill Duin's previous role) in strong collaboration with the University's Student Services office. Lynn's capabilities, in addition to providing advice and direction based on access to all advising documents related to these programs, would also include image recognition and emotional response to students. Algorithmic transparency would come through continued use of open access instructions and development methods. All structured and unstructured data from Lynn's interactions with students as well as personal information would be secure and safeguarded within university network and storage protocols. The Controller Office would be regularly involved to audit Lynn's technical system, and the program directors would regularly monitor its advising direction. We use this as a brief example; actual documentation of responses to this AIA questionnaire would result in significantly more detailed direction for attending to and mitigating risk. When putting it into professional practice, we need to inform the adoption of digital employees, like Lynn, through a process-driven approach to risk and accountability.

Strategic Approaches

Governments, institutions, and corporations face immense challenges surrounding the emergence of augmentation technologies and relentless AI development. Companies are able to design and deploy AI systems with substantial potential for harm or misuse, with governments largely and consistently being surprised (Whittlestone & Clark, 2021). Jess Whittlestone, at the University of Cambridge, and Jack Clark, from Anthropic (https://www.anthropic.com), an AI safety and research company, outline a proposal for improving the governance of AI by investing in greater government capacity for measuring and monitoring the capabilities and impacts of AI systems. They write that "if adopted, this would give governments greater information about the AI ecosystem, equipping them to more effectively direct AI development and deployment in the most societally and economically beneficial directions" (p. 3). Specifically, they propose that governments "invest in building infrastructure to systematically measure and monitor the capabilities and impacts of AI systems" (p. 5). On the basis of their research, they highlight eight policy challenges that pilot projects might address. These include identifying and mitigating biases in ML models; understanding how competitive a nation's AI ecosystem is; understanding progress in "policy-relevant AI domains" such as computer vision systems and smart industrial robots; accelerating commercial applications; understanding how AI advances alter jobs that people do; ensuring that policymakers are aware of potential harms of AI systems; understanding AI deployment in specific sectors; and identifying which actors policymakers will need to regulate. Governments are monitoring and forecasting technology emergence now with the same rigor as finance or climate change. TPCs are and will continue to be at the forefront of this dynamically changing domain.

European Union (EU) AI Act (AIA)

The European Union (EU) AI Act (AIA) works to establish the first comprehensive set of regulations for AI, setting a worldwide standard for augmentation technologies and AI. The proposal for an AI Act was presented in April 2022. It was developed in response to:

> explicit requests from the European Parliament (EP) and the European Council... for legislative action to ensure a well-functioning internal market for artificial intelligence systems ('AI systems') where both benefits and risks of AI are adequately addressed at Union level. It supports the objective of the Union being a global leader in the development of secure, trustworthy and ethical artificial intelligence as stated by the European Council and ensures the protection of ethical principles as specifically requested by the European Parliament.
>
> *(European Commission, 2021)*

Its four specific objectives are the following:

- ensure that AI systems placed on the Union market and used are safe and respect existing law on fundamental rights and Union values;
- ensure legal certainty to facilitate investment and innovation in AI;
- enhance governance and effective enforcement of existing law on fundamental rights and safety requirements applicable to AI systems; and
- facilitate the development of a single market for lawful, safe and trustworthy AI applications and prevent market fragmentation.

Its proposed rules will be enforced through governance at the Member States level, and great care has been taken to align this policy with other Union policies. The AI Act prohibits unacceptable AI practices and regulates high-risk AI systems, and it imposes transparency requirements to ensure that humans are aware when being exposed to AI. Products are subject to a third-party conformity assessment (EU product safety framework), and providers "must establish and document an appropriate post-market monitoring system to continuously check compliance with regulatory requirements." Application of the Regulation is to be ensured by a European Artificial Intelligence Board, and fines for non-compliance can be as high as EUR 30,000.00. Essentially, the European Commission ensures that its approach centers on excellence and trust: "The European approach to AI will ensure that any AI improvements are based on rules that safeguard the functioning of markets and the public sector, and people's safety and fundamental rights" (EURO Smart, 2022).

The US National Artificial Intelligence Research Resource (NAIRR)

In contrast, during this similar time period, the US Office of Science and Technology Policy (OSTP) and the National Science Foundation (NSF) formed the National Artificial Intelligence Research Resource Task Force (NAIRRTF, 2022) to develop a roadmap for spurring AI innovation through access to critical resources and tools. All meeting agendas, presentations, minutes, and recordings are available at https://www.ai.gov/nairrtf/. Of note in October 2021 are the 84 responses to the task force's initial report from government, academic, and industry stakeholders. Here we share responses from the first seven reports, as these exemplify overall responses to the initial report. We also include responses from entities referred to in this book's resources; these include Deloitte, IBM, IEEE, Microsoft, and Partnership on AI.

Overall, the corporate responses advance the need for sharing of data:

- Accenture corporation
  - "The primary goal for the NAIRR should be to serve as a champion for the sharing of data. The field of AI research would make tremendous

advances if private industry and government agencies shared more data and if needed protections and systems were in place to allow for such sharing to occur."

(p. 7)

- AI Redefined (AIR), a Montreal-based company building AI training for humans and machines
  - "An open-source orchestration platform enabling the access and combined use, in shared environments, of different AI agents, as well as human users, is of paramount importance for several reasons:
    - The ability to easily compare and audit implementations of AI agents with: other AI implementations, human users, other non-learning algorithms or heuristics, the ability to keep human oversight, [and] the ability to provide context to AI agents through human expertise."

(pp. 36–37)

- IBM
  - "The NAIRR must include the following core components:
    1   Federated, hybrid cloud enabled computing resource—an accessible and easy-to-use hybrid- and multi-cloud computing resource built on open architecture that amalgamates various public clouds (such as Amazon, Azure), private clouds, and on-premises resources to create a single, unified, flexible compute infrastructure.
    2   Data and models—large-scale, high-quality, trusted, AI-ready datasets and pre-trained AI models across the broad AI science and technology landscape.
    3   Software and tools—integrated and interoperable software and platform technologies that support AI research and development and enable those with varying degrees of technology and science expertise to be productive.
    4   Education—training materials, outreach activities, and user support that ensures easy, efficient, and effective use of the NAIRR."

(p. 309)

Academic AI institutes advise against pursuing shared research infrastructure:

- AI Now Institute of New York University and Data & Society Research Institute
  - "The NAIRR will entrench, rather than challenge, corporate control over the AI field."

(p. 25)

    - "We therefore advise that the NAIRR's forthcoming roadmap and implementation plans recommend against pursuing shared research infrastructure, and instead explore alternative ideas for expanding research into AI, increasing government investment in critical work on AI, and meaningfully democratizing decision-making on the development and deployment of AI and related technologies" (p. 28)

- "The NAIRR Task Force must reckon with mounting evidence of the harmful impacts of large-scale AI systems, including discriminatory consequences for marginalized groups and long-term climate impact" (p. 25).
- "The NAIRR, as presently conceived, bolsters misleading and dangerous 'tech cold war' narratives, which reflect the self-interest of Big Tech and the defense contracting industry, without being backed by robust evidence" (p. 25).
  - "Tech company CEOs and former national security officials have been some of the most vocal in endorsing this self-serving narrative, which troublingly echoes cold war framings that served in the past to accelerate government investment in weapons-related computing" (p. 32).

Non-profit organizations question the premise that the democratization of AI will lead to positive returns:

- American Civil Liberties Union (ACLU)
  - "The current framing of this resource takes the technochauvinist view that the most pressing problem with AI is its inaccessibility and that the democratization of AI will inevitably lead to positive returns for the United States. But expansions of AI that lack an express, ongoing focus on how AI development will affect civil rights and civil liberties will invariably lead to technologies that threaten these important protections."
  (p. 59)

Multi-stakeholder groups place a large emphasis on education and greater access and accountability:

- Alexandria Archive Institute (Open Context)
  - "Options for the goals for the establishment and sustainment of a NAIRR and metrics for success should include a heavy emphasis in education, demonstrable benefit to communities whose data are used to build the curated data sets used by the resource, an incorporation of inclusive governance for the resource, and improvement to infrastructure that allows easier access to communities currently underserved by AI research".

  (p. 41)
- American Psychological Association (APA)
  - "The Task Force has proposed several potential capabilities and services necessary to maintain a shared computing infrastructure and facilitate equitable access to resources for researchers across the country. Among

these capabilities, it is essential to prioritize the appropriate choice and use of metadata—this will enhance the understanding, organization, and use of curated datasets under NAIRR".

(pp. 70–71)

- Deloitte
  - "The TF [task force] will need to navigate: 1) need for openness & transparency to foster cutting-edge innovation while adhering to security demands of sensitive and proprietary data; 2) need to collaboratively address challenges of interoperability across data types and models; and 3) need to foster rapid innovation and R&D with a focus on ethical and trustworthy AI as well as diversification of AI developer talent."

(pp. 228–229)

- IEEE
  - "Partnership is essential to enhance credibility and ownership of the agreed approaches and ethical value preservation. At the same time, the use of public data by private parties and vice versa could pose problems if not embedded in a clear and ethical framework. Providing transparency, ensuring accountability, avoiding unfair bias, and offering the chance to opt out will be crucial."

(p. 322)

Responses across all sectors reiterate the importance of serving underrepresented and underserved communities:

- Amazon
  - "Increase the extent to which AI technology benefits underrepresented and underserved communities. AI technology has the potential to improve myriad critical products and services for all segments of society (e.g., legal, healthcare, financial services, etc.).
  - Increase the size and skills of the national AI workforce. To take full advantage of the opportunities that AI presents, we need to increase the number of people qualified to fill the new technical and non-technical jobs being created by the AI boom."

(p. 49)

- Microsoft
  - "Microsoft enthusiastically supports the aims of the National AI Research Resource (NAIRR) Task Force and shares the vision of democratizing access to the evolving ecosystem of foundational computing capabilities to empower a larger and more diverse artificial intelligence (AI) research and development (R&D) community."

(p. 410)

- Partnership on AI (PAI), a non-profit partnership of academic, civil society, industry, and media organizations

- "The National AI Research Resource should encourage stakeholders developing and using demographic datasets to:
  - Curate Datasets with support of community-based organizations that have trust and experience with the groups and communities in question.
  - Ensure the need for demographic data to assess anti-discrimination, fairness, and inequality does not infringe on privacy rights or increase the undue surveillance of protected classes, vulnerable populations, or marginalized groups.
  - Engage with the tensions around demographic data usage in AI and align with emerging work on the importance of equitable data."

(p. 538)

On January 24, 2023, the NAIRR Task Force submitted its final report to the U.S. President and Congress. Titled *Strengthening and Democratizing the U.S. Artificial Intelligence Innovation Ecosystem*, this report "proposes an implementation roadmap for the NAIRR that would build on existing and future Federal investments; design in protections for privacy, civil rights, and civil liberties; and promote diversity and equitable access." It also presents a phased approach for development of the "administration, policies, security framework, and cyberinfrastructure" needed for support of "the full spectrum" of AI research. While this report is US-centric, a global partnership ecosystem is needed for approaching governance and regulation surrounding the ethical development and deployment of augmentation technologies and AI.

NSF Artificial Intelligence Research Institutes

A specific strategic approach is the awarding of funds in support of AI research. As of 2022, the US National Science Foundation has awarded 27 Artificial Intelligence Research Institutes under six theme areas, including intelligent agents, neural and cognitive foundations, AI for agriculture and forestry, AI for decision-making, trustworthy AI, and AI-augmented learning. The National Artificial Intelligence (AI) Research Institutes program solicitation (2022) aligns with core statements throughout this book, namely that:

AI enables computers and other automated systems to perform tasks that have historically required human cognition and human decision-making abilities. Research in AI is therefore concerned with the understanding of the mechanisms underlying thought and intelligent behavior and their implementation in machines.... As intelligent systems amplify humans' capabilities to accomplish individual and collective goals, research is needed to assess the benefits, effects, and risks of AI-enabled computing systems; and to understand how human, technical, and contextual aspects of systems interact to shape those effects. Relevant research areas therefore include consideration of explainable and trustworthy AI; validation of AI-enabled systems; AI safety, security, and privacy; and the role of emotion and affect in the design and perception of increasingly sophisticated machine intelligence.

Of interest to practitioners is the Institute for Foundations of Machine Learning (https://www.ifml.institute), whose goal is to "create new algorithms that can help machines learn on the fly, change their expectations as they encounter people and objects in real life, and even bounce back from deliberate attempts by adversaries to manipulate datasets." One of the current research projects, Fairness in Imaging with Deep Learning, works to design metrics to ensure that deep learning imaging algorithms reflect the diversity of our world.

Of specific interest to instructors is the Institute for Student-AI Teaming (https://www.colorado.edu/research/ai-institute/), whose goal is to move toward a future "where AI is viewed as a social, collaborative partner that helps students work and learn more effectively, engagingly, and equitable, while helping educators focus on what they do best: inspiring and teaching students." This Institute will "develop, deploy, and study AI Partners that interact naturally with students and teachers through speech, gesture, gaze, and facial expression in real-world classrooms and remote learning settings."

US Agency for International Development AI Plan

Another strategic approach is the development of agency action plans. The US Agency for International Development (USAID) released its *Artificial Intelligence Action Plan* in May 2022 beginning with a clear statement that it is not intended to impose any legal commitments on the US: "It does not reflect the United States Government's official position on artificial intelligence; rather, it is intended to inform USAID's approach to responsibly engaging with artificial intelligence technologies." The Action Plan begins with acknowledgment of AI applications' projected impact on the global economy while at the same time discussing surveillance, censorship, cybersecurity, disinformation, and inequity and bias concerns. They advocate for the use of AI in USAID work "when its benefit is clear—but importantly, also advocating against its use when the risk of harm is too great." Regarding their approach, they emphasize responsible AI:

> a responsible approach to AI should include strengthening key aspects of the enabling ecosystem. This includes data systems, connectivity, and local workforce capacity. In addition, there must be a focus on strengthening the civil society structures holding AI systems and actors accountable,

and shaping policy environments that in turn encourage open, inclusive, and secure digital ecosystems.

(p. 1)

And the three concrete steps for responsible AI are the following:

*   Commit to Responsible AI in USAID programming;
*   Strengthen digital ecosystems to support responsible use of AI;
*   Partner to shape a global Responsible AI agenda (p. 2).

We also recommend use of the USAID document *Managing Machine Learning Projects in International Development: A Practical Guide* (2021) as an excellent introduction to AI and machine learning and guide when evaluating feasibility of use of AI as well as for design, implementation, and post implementation project management. While designed for "development practitioners who may not be trained technologists but are involved with or responsible for implementing projects that might have a technical machine learning/artificial intelligence component" (p. 4), readers will find the guidance to be useful.

US Food and Drug Administration (FDA)

As a final example in this section on strategic approaches, we touch on the future direction for regulating technologies that have medical or therapeutic uses as well as consumer applicability. Neurotechnologies are promising to either solve or overcome major health issues, but also to advance human limitations through enhancement. Cognitive enhancement augmentation technologies can be achieved through brain-computer interfaces (BCI) in the form of brain implants. They are designed "to serve therapeutic ends and thus are subject to FDA regulation" (Binkley et al., 2021). The United States Food and Drug Administration (FDA) is an agency of the Department of Health and Human Services at the federal level. Its mandate serves manifold activities; one is to enforce laws to protect consumers, regulating "$1 trillion worth of products a year" (FDA, 2021a). For example, the FDA has produced regulations and product classifications for ingestible technologies, which is one of the embodied platforms we track in Fabric of Digital Life (Iliadis, 2020). In 2021, the FDA's Center for Devices and Radiological Health (CDRH) published *Implanted Brain-Computer Interface (BCI) Devices for Patients with Paralysis or Amputation—Non-clinical Testing and Clinical Considerations: Guidance for Industry and Food and Drug Administration Staff* (FDA, 2021b). The document produces non-binding recommendations that do not include the force and effect of law. However, it provides guidance for those involved in the development of these devices including "stakeholders (e.g., manufacturers, health-care professionals, patients, patient advocates, academia, and other government agencies) [to] navigate the regulatory landscape for these medical devices" (p. 1).

The document describes brain implant devices for patients through basic criteria or "modules" (pp. 3–4). At a minimum, each would include most, but not necessarily all of these modules:

A   Signal acquisition (e.g., leads and recording electrodes);
B   Signal processing that includes software for decoding and encoding signals and providing stimulation (in some cases) and associated hardware;
C   Stimulation delivery (internal/external stimulator and stimulating electrodes);
D   Assistive effector component (e.g., a prosthetic limb, wheelchair, functional electrical stimulators applied to intact limbs, exoskeletons or robotic systems, or communication devices and computers);
E   Sensor component for neural feedback (e.g., sensors for restoring touch or reporting other information), if applicable; and
F   Programming module that consists of an operating protocol to control functions, such as turning the device on and off and switching between various outputs and programs (FDA, 2021b, pp. 3–4).

One can see how an augmentation technology is interpreted through this list of components and processes to achieve physical enhancement. Take for example the *assistive effector component*, which covers the way brain implants would be used to control another device, such as an exoskeleton or prosthetic limb. The document goes on to explain:

> In the case of software that will control various assistive effector components (i.e. motorized wheelchairs, computer software, upper limb prosthetics), we recommend that you account for any software-related hazards, and associated changes due to algorithm updates, in your risk-analysis plan
>
> *(p. 8)*

The function of technical communicators comes to the fore; they will be responsible for BCI documentation across these genres.

Health-related augmentation technologies are designed as consumer devices and are subsequently being adapted in cultural spheres. Charles E. Binkley, Michael S. Politz, and Brian P. Green (2021) explain that "BCIs will not simply augment a single person; they present the potential for a bifurcation between 'enhanced' and 'standard' human beings." They argue that the blurring of BCI usage for therapy and human enhancement (augmentation) makes FDA approval problematic:

Implantable brain-computer interface (BCI) and other devices with potential for both therapeutic purposes and human enhancement are being rapidly developed. The distinction between therapeutic and enhancement uses of these devices is not well defined. While the US Food and Drug Administration (FDA) rightly determines what is safe and effective, this article argues that the FDA should not make subjective, value-laden assessments about risks and benefits when it comes to approval of BCIs for therapy and enhancement.

Binkley, Politz, and Green's point is relevant to this book: that value-based judgments about enhancements should not necessarily be made by the FDA.

Examples of brain implants for consumer purposes, laden with transhumanist rhetoric, rage in popular media. In American contexts, it will be the FDA that ultimately polices the availability of these consumer products. While Neuralink is often heavily sensationalized by owner Elon Musk for brain enhancement, other companies are proposing brain enhancement for both therapeutic and consumer ends. Synchron Inc. received FDA approval in July 2021 to test brain implants on human subjects (McBride, 2021) and implanted several people in both Australia and the United States (Botros, 2022). Synchron device users are able "to send messages through WhatsApp and make online purchases" from their brains (Botros, 2022). In a recent TED talk, Synchron CEO and neurologist Tom Oxley discusses how brain implants might restore autonomy and dignity to those with paralysis. He also mentions the future of communication, and that it might be significantly transformed by this technology. He talks about how it could augment the way people express their emotions to others without using words, just feelings (Oxley, 2022).

Other aspects of brain-computer communication are underway. Standards have already been set to connect embodied technologies (implants and wearables) to external servers through the IEEE body-area network standard (IEEE 802.15.6 "BAN" standard). These body networks will send data from bodies and brains to external actors such as doctors, police, or insurance agents, helping to provide the wireless structure for significant advancements (Pedersen, 2020). However, as standards advance (such as the BAN standard), laws, regulations, and policies will also need to evolve for these biomedical devices before highly invasive technologies become consumer devices.

Ethical considerations are still outstanding and bioethicists have long questioned several issues. Paul R. Wolpe, writing in 2007, says of brain implants:

> they will challenge personal autonomy, as experiments with other animals show how the brain can be conditioned or even disrupted with implanted technologies. And they will challenge our conceptions of selfhood, when computers are part of the very functioning of our thought processes.
>
> (p. 130)

Further, bioethics should help better serve equity-seeking citizens as organizations such as the FDA begin to further regulate brain-computer interaction. Keisha Shantel Ray (2021) writes:

> Black bioethics considers how racism and white supremacy touch every aspect of the lived experience of being Black in America. This extends to medicine since medicine continues to impart the effects of racism and white supremacy to marginalized populations like Black people.
>
> *(p. 40)*

Tactical Approaches

Technical and professional communicators and their associated industries also face immense challenges surrounding the emergence of augmentation technologies and relentless AI development. For the design, deployment and use of augmentation technologies and AI, we suggest that TPCs assume a tactical mindset. To begin to understand the depth and breadth of augmentation technologies and AI, one might peruse the 397 artifacts that are currently part of the Fabric collection *Human-centered Design for Augmented Reality* (Pedersen, 2018). These artifacts "allow us to see how the landscape of immersive technologies is shifting and how the understanding of what our role as technical communicators in this landscape must also shift" (Armfield et al., 2019). Based on this ongoing curation as well as our collective work in designing, deploying, and studying AR and VR devices (Duin et al., 2020), we again emphasize that TPCs must engage in participatory design and understand that work with augmentation technologies and AI demands an experiential, tactical mindset. Where once TPCs were required to be objective and non-existent in the presentation of information and documentation, we now must integrate our own experiences tactically into and throughout the process. One means for doing so is through coalition building and participation in professional standards forums.

Coalition Building

Collaboration is a disciplinary assumption in TPC. Scholars, instructors, and practitioners grapple with ever-changing modes and models for collaborative work in academia and industry. Technical and professional communicators must be prepared to collaborate with engineers, subject matter experts, and programmers; they must be adept at using collaborative software and working with global virtual teams:

> William Duffy (2014), in his review of the decades of scholarship on collaboration, notes that Bruffee's 'conversational imperative' sets the stage

for what is known largely as the social constructivist epistemology, or the "social turn" in our larger discipline (p. 417).

(Duin et al., 2021, p. 172)

This social turn has served as a lasting lens within which rhetoricians theorize collaborative efforts. Moreover, open access and open collaborative tools invite "radical sharing," inviting TPCs to emerge from silo workspaces and tackle complex disciplinary and professional challenges. Moreover, TPCs now must be prepared to collaborate with digital employees.

In 2019, Rebecca Walton, Kristin Moore, and Natasha Jones in their seminal work, *Technical Communication after the Social Justice Turn: Building Coalitions for Action*, helped scholars, instructors, and practitioners to recognize "the injustices and oppressive systems embedded in our work as technical communicators" (p. 133). As a means to address these injustices, they state:

> At the heart of this book is the belief that technical communicators can and should build coalitions and that through intersectional, coalitional approaches to technical and professional communication (TPC), we can address issues of inequality and oppression. But we also believe that we need practical strategies and tactics for getting this work done.
>
> *(p. 133)*

Their four steps for "redressing inequities" are Recognize, Reveal, Reject, and Replace (4Rs):

- Recognizing injustices, systems of oppression, and our own complicities in them;
- Revealing these injustices, systemic oppressions, and complicities to others as a call-to-action and (organization/social/political) change;
- Rejecting injustices, systemic oppressions, and opportunities to perpetuate them;
- Replacing unjust and oppressive practices with intersectional, coalition-led practices (p. 133).

They advocate for "intersectional coalition building," an approach that acknowledges the different lived experiences of those who are marginalized, and for those who are not "living at the intersections of oppression," to approach change-making coalitions "with humility; to listen more than they speak or lead; and to sometimes divest themselves of self-serving plans, ideas, and ways forward" (p. 134). They emphasize that "the important thing about intersectional coalition building is that the right answer, the next step, is localized and should be driven by the collective agenda and the experiences of those who have been and continue to be multiply marginalized" (p. 134). Chapter 5 in this text concentrates on socio-ethical consequences of augmentation technologies and AI, elaborating on guiding principles for development and deployment. Throughout this book we have worked to emphasize the injustices and systemic oppressions that arise amid augmentation technologies and AI. While algorithmic impact assessments and other tools assist in providing opportunities for ameliorating injustices, most critical is to replace "oppressive practices with intersectional, coalition-led practices" (p. 133).

Standards Building

We also consider standards building to be a tactical opportunity for TPCs to participate in the ethical emergence of augmentation technologies across the four enhancement categories, sensory, cognitive, emotional and physical. While standards are not enforceable, they represent ethical, standardized practices familiar to TPC professionals. The International Standards Organization (ISO), for example, develops standards to ensure companies operate using a set of best practices. Following them helps companies gain and maintain international credibility. Another example well known to TPCs is the Darwin Information Typing Architecture (DITA) standards for defining a set of document types for authoring and organizing topic-oriented information. The OASIS DITA Technical Committee defines and maintains the DITA standards.

Augmentation technologies are slated to meet ambitious goals involving significant convergences and coalitions, all wrapped in an expectation for a transformed, next-stage internet. Using the metaphor of a constellation, the *Metaverse Standards Forum* was established in June of 2022. It states its goal, "building a pervasive, open and inclusive metaverse at a global scale will require cooperation and coordination between a constellation of international standards organizations" (https://metaverse-standards.org). This forum will not publish any standards; rather, it seeks to encourage standards bodies to develop open standards and to use them. Khronos Group's President Neil Trevett (who is also VP Developer Ecosystems at NVIDIA) discusses the need for interoperable standards in order to keep a check on proprietary technology:

> If there's a big lag between the technology becoming available and the standard that makes it openly available... then there's a danger that proprietary technologies are going to get baked into the infrastructure of the metaverse, and I don't think anyone really wants that.
>
> *(Ravenscraft, 2022)*

By looking at the member organizations, one can extrapolate their roles. Members include the following:

Khronos Group, https://www.khronosgroup.org

A non-profit consortium to develop open standards for 3D graphics, virtual and augmented reality, parallel computing, machine learning, and vision processing.

World Wide Web Consortium (W3C), https://www.w3.org

An international, non-profit community consortium to develop web standards, including HTML, CSS, SVG, Ajax, and other technologies for web applications.

Open Geospatial Consortium (OGC), https://www.ogc.org

A standards consortium to make geospatial (location) information and services findable, accessible, interoperable, and reusable.

OpenAR Cloud, https://www.openarcloud.org

A non-profit organization to drive the development of open and interoperable spatial computing technology, data, and standards, to connect the physical and digital worlds.

Spatial Web Foundation, https://spatialwebfoundation.org

An integrated, ethically aligned network, with the mission to develop and maintain the technical and ethical standards of the spatial web.

One major issue that these organizations need to solve is a persistent digital infrastructure, a connection between digital addresses and physical space because a Metaverse cannot emerge without one. This concept is dubbed the "AR cloud." A 2021 World Economic Forum Report, *Technology Futures: Projecting the Possible, Navigating What's Next*, describes the AR cloud in terms of its future potential:

> The AR cloud has the potency to launch a near-term revolution in the spatial web by mapping everything we do in the real world to digital information and enhancements. As one study puts it, the spatial web would "fully erase the line between digital and physical objects". Surgeons would, for example, perform diagnostics and surgery from another part of the world thanks to advanced haptics (technology that creates an experience of touch), precision robotics and enhanced 3D digital modeling, or students would learn about blood cells by exploring the human bloodstream virtually. The possibilities are endless.
>
> *(p. 25)*

The World Economic Forum points to cutting-edge ways that an AR cloud could help people.

In October 2021, just as Mark Zuckerberg announced META's intent concerning the Metaverse (2021), we formed a panel of researchers to report on the AR cloud. Our paper was titled "The AR Cloud: Tech Imaginaries, Future Risks, and Potential Affordances" for the Society for the Social Studies of Science conference with others presenting on Critical Augmented Reality Studies (Pedersen et al., 2021). We explained the AR cloud's significance:

> The AR cloud requires a persistent 3D digital copy of the world, which means that if a developer creates a virtual overlay for a real space, it will remain accessible to anyone after the fact across different devices and platforms. As an imagined technology future, many believe it could affect positive societal change (e.g., Open AR Cloud Association), while others believe the risk could involve further conditions of surveillance, corporate control, or harm to humans.

The OpenAR Cloud (2022) association has set up OARC Working Groups (WG) to include members to help "define and tackle the challenges facing the successful achievement of real world spatial computing usage every day and for the benefit of all of humanity." They cover issues such as ensuring open-source projects for the spatial computing sector to help evolve an ecosystem that is safe and accessible for all. In places, OpenAR asks pressing questions concerning ethical emergence like "Should we attempt to put up 'firewalls' around countries that respect human rights in spatial computing so that 'rogue' countries cannot abuse the Open AR Cloud ecosystem to violate the same human rights?"

Recall de Certeau's (1984) emphasis on a tactic as "a calculated action... [that] takes advantage of 'opportunities' and [that] vigilantly make[s] use of the cracks that particular conjunctions open in the surveillance of the proprietary powers" (p. 37). He also writes that "the space of a tactic is the space of the other. Thus it must play on and with a terrain imposed on it and organized by the law of a foreign power" (p. 37). Augmentation technologies and AI may well represent a "foreign" terrain imposed on society and our TPC work through exponential algorithmic development, deep and machine learning, and emerging digital employees. Each of the approaches discussed in this chapter includes opportunities for advancing ethical futures for augmentation technologies and AI in technical communication, for engaging with this dynamic phenomenon. Practitioners, instructors, scholars, and students can use the outstanding reports noted earlier as a means to build awareness of technology trends, map ethical and human rights-based approaches to augmentation technologies and AI, practice and integrate use of algorithmic impact assessments in project development, promote societal and policy guidelines in support of human-centric systems, and address and ameliorate risk surrounding gender, race, and power in AI.

Conclusion

The purpose of this book is to cultivate an even deeper understanding of human augmentation and AI technology, build TPC capacity to articulate its benefits and risks, and provide direction for future practice and collaboration. Augmentation technologies fueled by AI will continue to undergo a process of normalization geared to everyday users in various roles as practitioners, caregivers, artists, educators, and students. The meaning of the term "workforce" will continue to evolve to include and integrate autonomous agents and digital employees with whom TPCs will collaborate within the AI ecosystem. Augmentation technologies have moved beyond amplifying capabilities to augmented, expanded human capability. As Zizi Papacharissi (2019) states, "We will, or rather, we have the opportunity to, become different" (p. 7).

Throughout this book we have worked to redefine augmentation technologies and AI as social and rhetorical phenomena. We have brought theoretical, empirical, pedagogical, critical, and ethical attention to the development of these sophisticated, emergent, and embodied augmentation technologies slated to promote enhanced futures—i.e., to improve lives, literacy, cultures, arts, economies, and social and professional contexts. We have worked to reframe professional practice and pedagogy to promote digital and AI literacy surrounding the ethical design, adoption, and adaptation of augmentation technologies. Only by understanding and embedding ethical principles throughout augmentation technology design and AI can we foster human-centered, humane futures. With this book's direction, our goal is that TPC scholars and practitioners will be positioned to assist in meeting the critical need for the ethical design, adoption, and adaptation of augmentation technologies and AI. Indeed, technical and professional communicators are best positioned to address the ethical and pragmatic challenges emerging.

We invite scholars, instructors, practitioners, and students to collaborate on research underway as part of the Digital Life Institute at https://www.digitallife.org. This international research network of multidisciplinary scholars studies the social implications of disruptive digital technologies. To date, five research clusters have formed to investigate digital life: AI in Education; AI Implications; Building Digital Literacy; Digital Cultural Heritage; and Sustainability, Equity, and Digital Culture. Feel free to contact us to share your interest in collaborating to investigate the human and social dimensions of digital technologies.

References

Angeli, E. L. (2019). *Rhetorical work in emergency medical services: Communicating in the unpredictable workplace.* Routledge.

Armfield, D., Duin, A. H., & Pedersen, I. (2019). Immersive content in technical communication: A journey mindset. *Intercom.* https://www.stc.org/intercom/

Bensinger, G. (2022, May 30). How Illinois is winning in the fight against big tech. *New York Times.* https://www.nytimes.com/2022/05/30/opinion/illinois-biometric-data-privacy.html.

Binkley, C. E., Politz, M. S., & Green, B. P. (2021, September). Who, if not the FDA, should regulate implantable brain-computer devices? *AMA Journal of Ethics*. https://journalofethics.ama-assn.org/article/who-if-not-fda-should-regulate-implantable-brain-computer-interface-devices/2021-09

Botros, A. (2022, July 18). Elon Musk's Neuralink brain computer startup is beat again. This time a competitor implanted its device into its first U.S. patient. *Fortune*. https://fortune.com/2022/07/18/elon-musk-neuralink-beat-by-synchron-brain-computer-startup-us-human-trial/

de Certeau, M. (1984). *The practice of everyday life*. University of California Press.

Ding, H. (2014). *Rhetoric of a global epidemic: Transcultural communication about SARS*. Southern Illinois University Press.

Ding, H. (2022). Foreword. In S. Stinson & M. Le Rouge (Eds.), *Embodied environmental risk in technical communication* (pp. xxiii–xxviii). Taylor & Francis.

Duffy, W. (2014). Collaboration (in) theory: Reworking the social turn's conversational imperative. *College English, 76*(5), 416–435.

Duin, A. H., Armfield, D., & Pedersen, I. (2020). Human-centered content design in augmented reality. In G. Getto, N. Franklin, S. Ruszkiewicz, & J. Labriola (Eds.), *Context is everything: Teaching content strategy* (pp. 89–116). Taylor & Francis.

Duin, A. H., & Pedersen, I. (2023). *Strategic and tactical approaches to designing ethical futures for augmentation technologies and AI* [Multimedia collection]. Fabric of Digital Life. https://fabricofdigitallife.com/Browse/objects/facets/collection:77

Duin, A. H., Tham, J., & Pedersen, I. (2021). The rhetoric, science, and technology of 21st century collaboration. In M. Klein (Ed.), *Effective teaching of technical communication: Theory, practice and application* (pp. 169–192). WAC Clearinghouse: Foundations and Innovations in Technical and Professional Communication series.

EURO Smart. (2022). *Artificial intelligence (AI) taskforce*. https://www.eurosmart.com/committees-and-task-forces/artificial-intelligence-ai/

European Commission (2020). *Welcome to the ALTAI portal!* https://futurium.ec.europa.eu/en/european-ai-alliance/pages/altai-assessment-list-trustworthy-artificial-intelligence

European Commission (2021, April 21). *Laying down harmonised rules on artificial intelligence (Artificial Intelligence Act) and amending certain union legislative acts*. https://eur-lex.europa.eu/legal-content/EN/TXT/HTML/?uri=CELEX:52021PC0206&from=EN

European Union's General Data Protection Regulation (GDPR) program (2022). https://gdpr.eu/

FDA (2021a). *Fact sheet: FDA at a glance*. US Food & Drug Administration. https://www.fda.gov/about-fda/fda-basics/fact-sheet-fda-glance

FDA (2021b). *Implanted brain-computer interface (BCI) devices for patients with paralysis or amputation—non-clinical testing and clinical considerations: Guidance for industry and Food and Drug Administration staff*. FDA Center for Devices and Radiological Health. https://www.fda.gov/regulatory-information/search-fda-guidance-documents/implanted-brain-computer-interface-bci-devices-patients-paralysis-or-amputation-non-clinical-testing

Fjeld, J., Achten, N., Hilligoss, H., Nagy, A., & Srikumar, M. (2020). *Principled artificial intelligence: Mapping consensus in ethical and rights-based approaches to principles for AI*. Berkman Klein Center Research Publication. https://ssrn.com/abstract=3518482

Fjeld, J., & Nagy, A. (2020). *Principled artificial intelligence: Mapping consensus in ethical and rights-based approaches to principles for AI*. Berkman Klein Center for Internet & Society at Harvard University. https://cyber.harvard.edu/publication/2020/principled-ai

Frost, E. A. (2013). Transcultural risk communication on Dauphin Island: An analysis of ironically located responses to the Deepwater Horizon disaster. *Technical Communication Quarterly, 22*(1), 50–66. https://doi.org/10.1080/10572252.2013.726483

Gallagher, J. R. (2017). Writing for algorithmic audiences. *Computers and Composition, 45*, 25–35.

Giles, T. D. (2010). Communicating the risk of scientific research. *Journal of Technical Writing and Communication, 40*(3), 265–281.

Government of Canada (2022). *Algorithmic impact assessment tool.* https://www.canada.ca/en/government/system/digital-government/digital-government-innovations/responsible-use-ai/algorithmic-impact-assessment.html

Hao, K., & Freischlad, N. (2022, April 21). The gig workers fighting back against the algorithms. *MIT Technology Review.* https://www.technologyreview.com/2022/04/21/1050381/the-gig-workers-fighting-back-against-the-algorithms/

Herndl, C. G., & Brown, S. C. (1996). *Green culture: environmental rhetoric in contemporary America.* University of Wisconsin Press.

Hill, K. (2020, January 18). The secretive company that might end privacy as we know it. *New York Times.* https://www.nytimes.com/2020/01/18/technology/clearview-privacy-facial-recognition.html

The IEEE Global Initiative on Ethics of Autonomous and Intelligent Systems (2019). *Ethically aligned design: A vision for prioritizing human well-being with autonomous and intelligent systems, First Edition.* IEEE Advancing Technology for Humanity. https://standards.ieee.org/content/dam/ieee-standards/standards/web/documents/other/ead1e.pdf

Iliadis, A. (2020, November). *Policy and privacy implications of ingestibles* [Multimedia collection]. Fabric of Digital Life. https://fabricofdigitallife.com/Browse/objects/facet/collection/id/49

Kak, A. (2020). *Regulating biometrics: Global approaches and urgent questions.* AI Now Institute. https://ainowinstitute.org/regulatingbiometrics.pdf

Kimball, M. (2017). Tactical technical communication. *Technical Communication Quarterly, 26*(1), 1–7. https://doi.org/10.1080/10572252.2017.1259428

McBride, S. (2021, July 28). *NYC brain computer startup announces FDA trial before Elon Musk.* Bloomberg.com. https://www.bloomberg.com/news/articles/2021-07-28/elon-musk-neuralink-competitor-announces-fda-trial-for-brain-device

Metaverse Standards Forum (2022). https://metaverse-standards.org/

Mirel, B. (1994). Debating nuclear energy: Theories of risk and purposes of communication. *Technical Communication Quarterly, 3*(1), 41–65. https://doi.org/10.1080/10572259409364557

Moss, E., Watkins, E. A., Singh, R., Elish, M. C., & Metcalf, J. (2021) *Assembling accountability: Algorithmic impact assessment for the public interest.* Data and Society Report. https://datasociety.net/library/assembling-accountability-algorithmic-impact-assessment-for-the-public-interest/

National Artificial Intelligence (AI) Research Institutes (2022). Program solicitation NSF 22–502. National Science Foundation. https://www.nsf.gov/pubs/2022/nsf22502/nsf22502.htm

The National Artificial Intelligence Research Resource Task Force (NAIRRTF) (2022). US Government. https://www.ai.gov/nairrtf/

The National Artificial Intelligence Research Resource Task Force (NAIRRTF) (2023, January 24). *Strengthening and democratizing the U.S. artificial intelligence innovation ecosystem: An implementation plan for a national artificial intelligence research resource.* https://www.ai.gov/wp-content/uploads/2023/01/NAIRR-TF-Final-Report-2023.pdf

Newman, J., Cleaveland, A., Gordon, G., & Weber, S. (2020). *Designing risk communications: A roadmap for digital platforms.* UC Berkeley, Center for Long-Term Cybersecurity. https://cltc.berkeley.edu/wp-content/uploads/2020/12/Designing_Risk_Communications.pdf

OpenAR Cloud. (2022). *Working groups.* https://www.openarcloud.org/workinggroups/overview

Ovide, S. (2022, June 16). The hands-off tech era is over. *New York Times.* https://www.nytimes.com/2022/06/15/technology/government-intervention-tech.html

Oxley, T. (2022). *A brain implant that turns your thoughts into text* [Video]. TEDConferences. https://www.ted.com/talks/tom_oxley_a_brain_implant_that_turns_your_thoughts_into_text/transcript?language=en

Papacharissi, Z. (Ed.). (2019). *A networked self and human augmentics, artificial intelligence, sentience.* Taylor & Francis.

Pedersen, I. (2018, March). *Human-centred design for augmented reality (AR)* [Multimedia collection]. Fabric of Digital Life. https://fabricofdigitallife.com/Browse/objects/facet/collection/id/20

Pedersen, I. (2019). *Multinational corporations, AI, and geopolitical influence: A policy brief.* Geopolitics of AI Symposium, Global Affairs Canada, March 12, 2019.

Pedersen, I. (2020). Will the body become a platform? Body networks, datafied bodies, and AI futures. In I. Pedersen & A. Iliadis (Eds.), *Embodied computing: Wearables, implantables, embeddables, ingestibles* (pp. 21–47). MIT Press.

Pedersen, I., Duin, A. H., Iliadis, A., & Efrat, L. (2021). *The AR cloud: Tech imaginaries, future risks, and potential affordances.* Society for Social Studies of Science (4S) Annual Conference 2021, October 6–9, 2021, Toronto.

Potts, L. (2013). *Social media in disaster response: How experience architects can build for participation.* Routledge.

Ravenscraft, E. (2022, July 3). What, exactly, is the Metaverse Standards Forum creating? *Wired.* https://www.wired.com/story/metaverse-standards-forum-explained/

Ray, K. S. (2021). It's time for a black bioethics. *American Journal of Bioethics, 21*(2), 38–40. https://doi.org/10.1080/15265161.2020.1861381

Reisman, D., Schultz, J., Crawford, K., & Whittaker, M. (2018). *Algorithmic impact assessments: A practical framework for public agency accountability.* AI Now Institute. https://ainowinstitute.org/aiareport2018.pdf

Sauer, B. A. (2003). *The rhetoric of risk: Technical documentation in hazardous environments.* Lawrence Erlbaum.

Sloane, M. (2020, August 25). Participation-washing could be the next dangerous fad in machine learning. *MIT Technology Review.* https://www.technologyreview.com/2020/08/25/1007589/participation-washing-ai-trends-opinion-machine-learning/

Stephens, M. (2022). *Metaverse and its governance.* The IEEE Global Initiative on Ethics of Extended Reality (XR). https://standards.ieee.org/wp-content/uploads/2022/06/XR_Metaverse_Governance.pdf

Stinson, S., & Le Rouge, M. (Eds.). (2022). *Embodied environmental risk in technical communication.* Taylor & Francis.

USAID (2021). *Managing machine learning projects in international development: A practical guide.* US Agency for International Development. https://www.usaid.gov/sites/default/files/documents/Vital_Wave_USAID-AIML-FieldGuide_FINAL_VERSION_1.pdf

USAID (2022). *Artificial intelligence action plan*. US Agency for International Development. https://www.usaid.gov/sites/default/files/documents/USAID_Artificial_Intelligence_Action_Plan.pdf

Wachter, S., Mittelstadt, B., & Russell, C. (2018). Counterfactual explanations without opening the black box: Automated decisions and the GDPR. *Harvard Journal of Law Technology, 31*(2), 1–52.

Waddell, C. (1996). Saving the great lakes: Public participation in environmental policy. In C. G. Herndl & S. C. Brown (Eds.), *Green culture: Environmental rhetoric in contemporary America* (pp. 141–165). University of Wisconsin Press.

Walker, K. C. (2016). Mapping the contours of translation: Visualized un/certainties in the ozone hole controversy. *Technical Communication Quarterly, 25*(2), 104–120.

Walton, R., Moore, K., & Jones, N. (2019). *Technical communication after the social justice turn: Building coalitions for action*. Taylor & Francis.

West, S. M. (2019). *Discriminating systems: Gender, race and power in AI*. AI Now Institute. https://ainowinstitute.org/discriminatingsystems.pdf

Whittaker, M., Alper, M., Bennett, C. L., Hendren, S., Kaziunas, L., Mills, M., Morris, M. R., Rankin, J., Rogers, E., Salas, M., & West, S. M. (2019). *Disability, bias, and AI*. AI Now Institute. https://ainowinstitute.org/disabilitybiasai-2019.pdf

Whittlestone, J., & Clark, J. (2021). *Why and how governments should monitor AI development*. University of Cambridge, Centre for the Study of Existential Risk. https://arxiv.org/pdf/2108.12427.pdf

WIPO (2019). *WIPO Technology Trends 2019: Artificial Intelligence*. World Intellectual Property Organization. https://www.wipo.int/edocs/pubdocs/en/wipo_pub_1055.pdf

WIPO (2021). *WIPO technology trends 2021: Assistive technology*. World Intellectual Property Organization. https://www.wipo.int/edocs/pubdocs/en/wipo_pub_1055_2021.pdf

Wolpe, P. R. (2007). Ethical and social challenges of brain-computer interfaces. *AMA Journal of Ethics, 9*(2), 128–131. https://doi.org/10.1001/virtualmentor.2007.9.2.msoc1-0702

World Economic Forum (2021, April). *Technology futures: Projecting the possible, navigating what's next*. https://www3.weforum.org/docs/WEF_Technology_Futures_GTGS_2021.pdf

Zuckerberg, M. (2021, October 28). *Meta: Founder's letter, 2021*. https://about.fb.com/news/2021/10/founders-letter/

INDEX

Note: **Bold** page numbers refer to tables and *italic* page numbers refer to figures.